The Translational Apparatus of Photosynthetic Organelles

NATO ASI Series

Advanced Science Institutes Series

A series presenting the results of activities sponsored by the NATO Science Committee, which aims at the dissemination of advanced scientific and technological knowledge, with a view to strengthening links between scientific communities.

The Series is published by an international board of publishers in conjunction with the NATO Scientific Affairs Division

A Life Sciences B Physics	Plenum Publishing Corporation London and New York
C Mathematical and Physical Sciences D Behavioural and Social Sciences E Applied Sciences	Kluwer Academic Publishers Dordrecht, Boston and London
F Computer and Systems Sciences G Ecological Sciences H Cell Biology I Global Environmental Change	Springer-Verlag Berlin Heidelberg New York London Paris Tokyo Hong Kong Barcelona Budapest

NATO-PCO DATABASE

The electronic index to the NATO ASI Series provides full bibliographical references (with keywords and/or abstracts) to more than 30 000 contributions from international scientists published in all sections of the NATO ASI Series. Access to the NATO-PCO DATABASE is possible in two ways:

- via online FILE 128 (NATO-PCO DATABASE) hosted by ESRIN, Via Galileo Galilei, I-00044 Frascati, Italy.
- via CD-ROM "NATO-PCO DATABASE" with user-friendly retrieval software in English, French and German (© WTV GmbH and DATAWARE Technologies Inc. 1989).

The CD-ROM can be ordered through any member of the Board of Publishers or through NATO-PCO, Overijse, Belgium.

Series H: Cell Biology Vol. 55

The Translational Apparatus of Photosynthetic Organelles

Edited by

R. Mache

Laboratoire de Biologie Moléculaire Végétale
Université Joseph Fourier, B. P. 53X
38041 Grenoble cedex, France

E. Stutz

Université de Neuchâtel
Laboratoire de Biochimie Végétale
CH-2007 Neuchâtel, Switzerland

A. R. Subramanian

Max-Planck-Institut für Molekulare Genetik
Ihnestrasse 73
W-1000 Berlin 33, FRG

Springer-Verlag
Berlin Heidelberg New York London Paris Tokyo
Hong Kong Barcelona Budapest
Published in cooperation with NATO Scientific Affairs Division

Proceedings of the NATO Advanced Research Workshop on The Translational Apparatus of Photosynthetic Organelles held in Grenoble, France, July, 3–6, 1990

ISBN-13: 978-3-642-75147-9 e-ISBN-13: 978-3-642-75145-5
DOI: 10.1007/978-3-642-75145-5

Softcover reprint of the hardcover 1st edition 1991

31/3140-543210 – Printed on acid-free-paper

PREFACE

Structure and function of chloroplast and nuclear genes related to the synthesis and function of the photosynthetic apparatus are topics regularly dealt with in workshops and symposia. One often neglects the fact that chloroplast specific ribosomes are crucially involved in the synthesis of these proteins and that chloroplast ribosomes are themselves of bipartite genetic origin. More recently considerable progress has been made in characterizing both chloroplast and nuclear genes coding for chloroplast ribosomal components. These results allow now to more precisely address older questions concerning evolution, biosynthesis and function of chloroplast ribosomes.

The organizers felt that time had come to meet and exchange ideas and data concerning the "Translational apparatus of photosynthetic organelles". Thus a NATO Advanced Research Workshop was held in Grenoble, France, which was attended by about 50 scientists and students from USA, Europe and Japan.

The organizers invited as key-note speaker H.G. Wittman from the Max-Planck-Institute in Berlin, but his too early death prevented this wish to become true. Instead, R. Brimacombe from the same Institute gave the introductory lecture on recent studies of *E. coli* ribosomes to set the tone for the intimate link between research done with bacterial and chloroplast ribosomes. The papers presented in this book overview domains of current interest and stress the fact that biosynthesis and function of the translational apparatus draws upon two separate genetic systems and therefore brings about particular regulatory problems with fundamental significance. The evolutionary kinship of chloroplast and prokaryotic ribosomes became evident once more but also some important differences between them.

The meeting was organized with the financial assistance of NATO. We thank the Direction of the Scientific Affairs Division for their help and support and all the participants for making this workshop a big success. Last but not least we express our gratitude to CNRS, the University J. Fourier in Grenoble and the City of Grenoble for their generous support.

Régis Mache, Grenoble, France
Erhard Stutz, Neuchâtel, Switzerland
Alap Subramanian, Berlin, Germany

CONTENTS

PLASTID rRNAs AND tRNAs : STRUCTURE, ORGANIZATION AND EXPRESSION OF THEIR GENES

PLASTID RIBOSOMAL PROTEINS AND THEIR GENES IN HIGHER PLANTS

A. Chloroplast Encoded Genes

B. Nuclear Encoded Genes

PLASTID RIBOSOMES AND FACTORS OF ALGAE AND OF CYANELLE

FUNCTIONAL ASPECTS

PHYSIOLOGICAL ASPECTS

STRUCTURE AND EXPRESSION OF rRNA GENES

H. Kössel
Institut für Biologie
Universität Freiburg
Schänzle-Straße 1
D-7800 Freiburg, FRG

Introduction

The translation process within photosynthetic organelles of higher plants and algae is mediated by ribosomes of the 70S type first described by Lyttelton (1960), which are different from the cytoplasmic 80S and the mitochondrial 75S ribosomes. In view of the key function of 70S ribosomes for the expression of plastome encoded genes and because of the phylogenetic relevance of ribosomal RNA structures and their genes, structural and functional analysis of chloroplast rRNA genes from various species of higher plants and algae have been carried out in several laboratories since the beginning of plant molecular biology at the end of the seventies (for reviews see Palmer, 1985; Gruissem, 1989; Sugiura, 1989; Steinmetz and Weil, 1989; Delp and Kössel, 1991). In this article a survey is given first of the major characteristics of chloroplast operon structures and their constituents and second of our present knowledge of various steps of rRNA operon expression.
The final section summarizes a general two state model of the small ribosmal subunit which was recently proposed on the basis of highly conserved complementary sequences between the 5'- and 3'- terminal regions of small ribosomal subunit RNA (Kössel et al., 1990).

NATO ASI Series, Vol. H 55
The Translational Apparatus of Photosynthetic Organelles
Edited by R. Mache et al.

A. <u>Structure of Chloroplast rRNA Operons</u>

A description of chloroplast rRNA operons from various species and their comparison either between themselves or with the rRNA operons from bacteria or from other compartments of eukaryotic cells can be made at the following structural levels:

1. <u>Number and arrangement of operon copies per DNA molecule</u>
Chloroplast DNAs from the majority of higher plants contain rRNA genes within a pair of inverted repeat regions (Fig. 1a). Each inverted repeat includes a complete rRNA operon consisting of genes coding for 16S, 23S, 4.5S and 5S rRNAs. However, with certain members of the legume family and the conifers notablė exceptions from the inverted repeat situation have been found (Fig. 1b). In the chloroplast DNA of these species no inverted

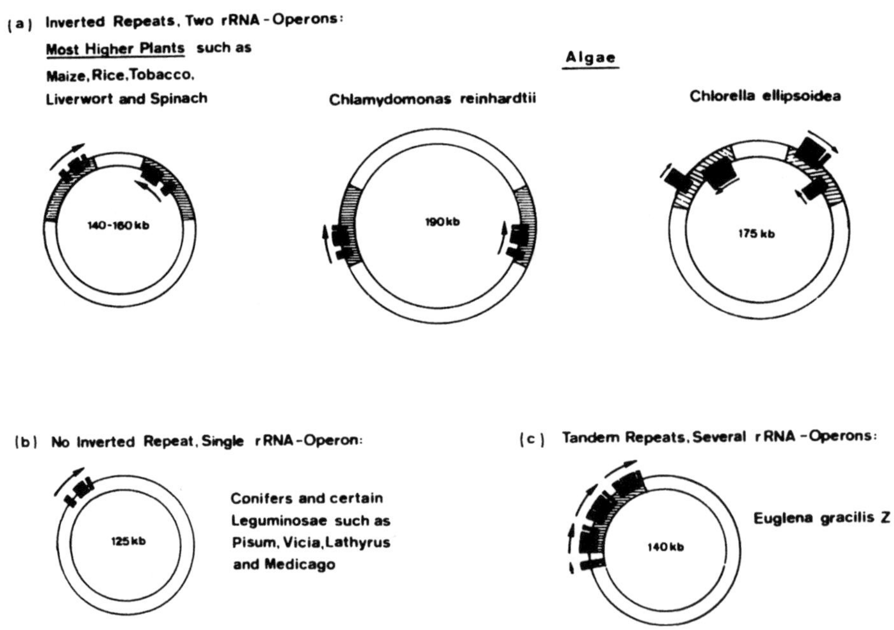

Figure 1. Repeat regions and rRNA operons on chloroplast genomes. Repeat regions are marked by shading. rRNA genes are symbolized by the black segments corresponding to 16S rDNA (middle sized segments), 23S rDNA (large segments) and 5S rDNA (small segments) with the arrows indicating the polarity of transcription. Note that the segment sizes are only approximately drawn to scale.

repeat and consequently only a single copy of an rRNA operon is present. A further exception is the alga Euglena gr. which contains three copies of complete rRNA operons in tandem repeat and in certain strains one extra copy of a 16S rRNA gene (Fig. 1c). On the other hand, chloroplast DNAs of other algae such as Chlamydomonas r. and Chlorella e. contain two rRNA operons within each of two inverted repeat regions (Fig. 1a) and therefore resemble the situation of most higher plants.

2. Nature and relative positions of individual components within the rRNA operons and their flanking regions

Most of the chloroplast rRNA operons show the typically prokaryotic gene order of 16S, 23S and 5S rDNA with a transcribed spacer between the 16S and 23S rRNA genes which encodes the two spacer $tRNA^{Ile}$ and $tRNA^{Ala}$ genes (Fig. 2). The only exceptions are the single 16S rRNA gene in Euglena (see Fig. 1c) and the rRNA operons of chlorella, where the second half of the operon

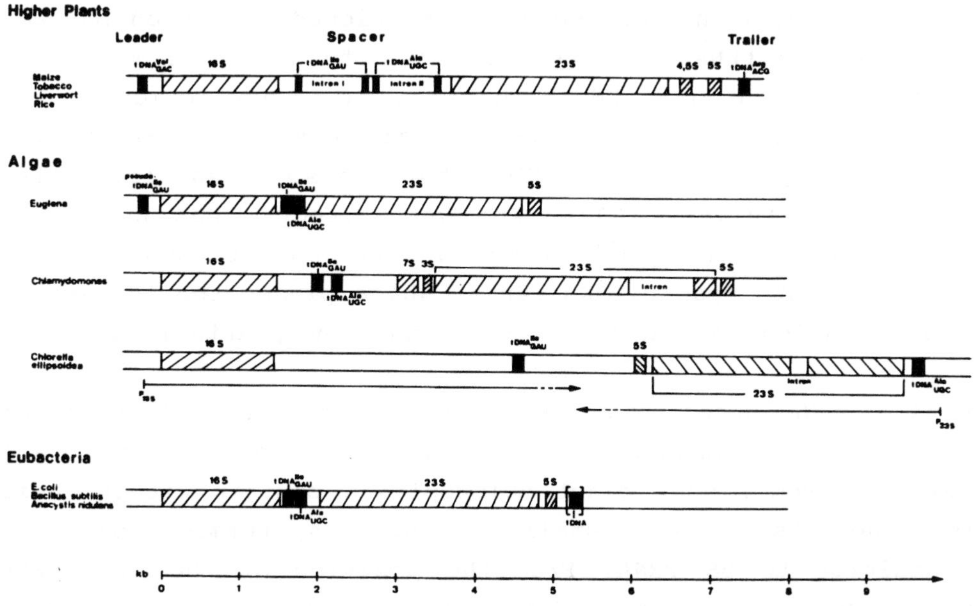

Figure 2. Alignment of rRNA operons from chloroplast of higher plants and algae and from eubacteria. Genes coding for rRNAs and tRNAs are marked by shading and black segments, respectively. In the case of Chlorella e. the two operons and their polarities are indicated by the long arrows starting at the respective promoter sites P_{16S} and P_{23S}.

which contains the spacer $tRNA^{Ala}$ and the 23S rRNA genes is inverted (Fig. 2). The latter situation results in two closely spaced but oppositely oriented operons contained in each of the inverted repeat region of the chloroplast DNA (Yamada and Shimaji 1987, see Fig. 1c).

Superimposed on the prokaryotic skeleton are elements characteristic for higher plant and algal rRNA operons, respectively. Thus as indicated in Fig. 2 the region proximal to the 16S rRNA gene contains a $tRNA^{Val}$ coding gene in all higher plants whereas Euglena 16S rDNA is flanked by a pseudo $tRNA^{Ile}$ gene on its 5' side. The spacer rDNA from all higher plant chloroplasts due to large intervening sequences in the tRNA genes are much longer as compared to their bacterial counterparts which contain the same spacer tRNA genes in an uninterrupted version. Intronless $tRNA^{Ile}$ and $tRNA^{Ala}$ genes are also found in the spacers of most of the algal RNA operons (Graf et al., 1980; Rochaix, 1987) including Chlorella (Yamada and Shimaji, 1987) where the spacer due to the above mentioned inversion between the two spacer tRNAs has been converted to a trailer region of 16S rDNA containing a $tRNa^{Ile}$ gene and a leader region of 23S rDNA containing the $tRNA^{Ala}$ gene. More recently algal species containing split spacer tRNAs have also been identified (Manhart and Palmer, 1990). Whereas Spirogyra has an intron only in the spacer $tRNA^{Ala}$ gene, Nitella and Coleochaete have introns in both spacer tRNA genes, which places the latter two families phylogenetically closer to the land plants (Fig. 3).

The 23S rRNA genes from chloroplasts show a phylogenetical tendency of fragmentation at their ends. For instance, all ribosomes from higher plant chloroplasts contain within their large subunits an extra small rRNA species, termed 4.5S rRNA, in addition to 5S rRNA. From the position of the 4.5S rRNA genes within the operons and from sequence homology it is clear that 4.5S rRNA is a structural equivalent of the 3'-terminal region of prokaryotic 23S rRNA. This 4.5S rRNA is, however, not present in chloroplast ribosomes from algae, liverworts, mosses and the fern Adiantum sp., but is again present in chloroplast

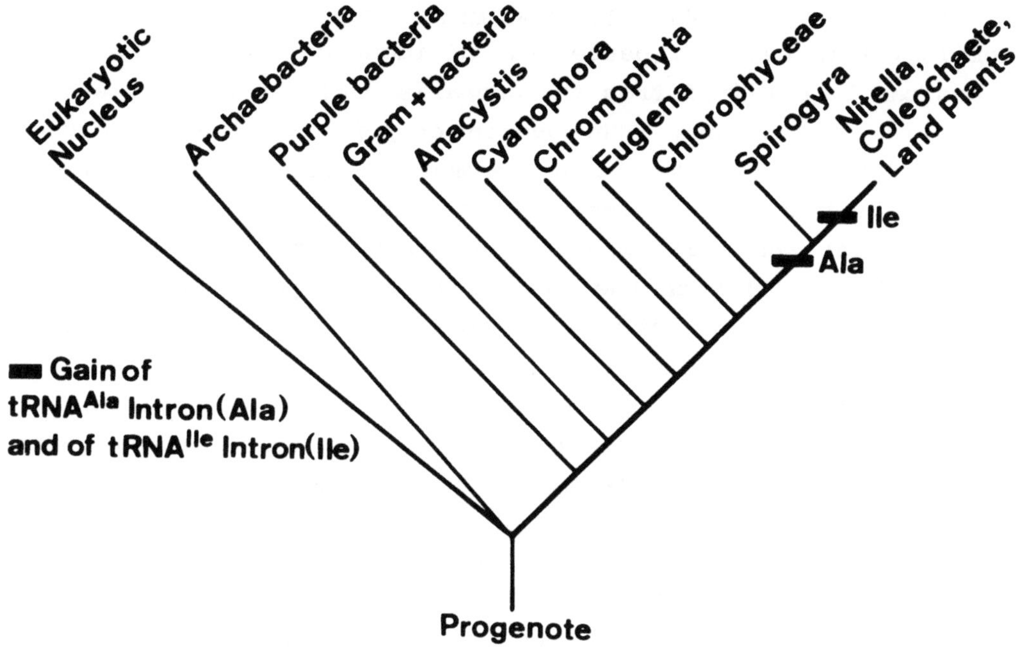

Figure 3. Cladogram based on the distribution of introns in the spacer tDNAs coding for tRNA^{Ile} and tRNA^{Ala} (according to Manhart and Palmer, 1990).

ribosomes from the fern Dryopteris acuminata. In ribosomes from Chlamydomonas chloroplast other small rRNAs, a 3S and a 7S rRNA, are found. Again positional and sequence homology of the corresponding genes show that they represent equivalents of 5'-terminal regions of prokaryotic 23S rRNA (Rochaix 1987). It should be emphasized that all the small rRNAs are contained in the primary transcripts in which they are therefore still connected with 23S rRNA in the form of a precursor rRNA. Thus fragmentation of plastid 23S rRNA genes occurs at the rRNA precursor level and is probably signalled by the sequences of the respective intergenic regions introduced during evolution.
Contrary to these fragmentations, the products of which remain unligated in the mature large ribosomal subunits, true introns are also observed in 23S rDNAs from Chlamydomonas and Chlorella (Fig. 2). As a prokarotic trait, 5S rRNA genes and in higher plants a trailer tRNA^{Arg} gene are positioned at the distal ends of chloroplast rRNA operons.

3. <u>Primary and secondary structures of rRNA transcripts</u>
Homologies in terms of percentage of identical nucleotide positions can be obtained from the comparison of the primary structures of the various rRNA operon components. The structural genes (rDNAs and tDNAs) show maximum homology (> 92%) between higher plant species, somewhat less homology (80-90%) between higher plant and algal species and even less but still strong homology (65-77%) with the bacterial counterparts. Of particular interest is the strong homology between the 3'-terminal region of the 16S rRNAs from chloroplasts and bacteria, as this supports an mRNA initiator site recognition by the chloroplast 30S subunit similar to the one proposed by Shine and Dalgarno for bacterial systems (Ruf and Kössel, 1988). Altogether a prokaryotic nature of chloroplast rRNA and tRNA genes is clearly evident from the sequence comparisons (Schwarz and Kössel, 1980; Graf et al., 1980; Koch et al., 1981; Schwarz et al., 1981, Edwards and Kössel, 1981; Graf et al., 1982, Pace et al., 1986). On the other hand leader regions (Przybyl et al., 1984), intergenic regions and intervening sequences show considerably less sequence conservation even between higher plant species and no homologies are detectable between leader or spacer sequences from higher plant and algal chloroplasts and bacteria (Koch et al., 1981; Graf et al., 1980).

Divergence of the rRNA genes is not only due to single base substitutions. Besides single base deletions (or insertions) which are often compensated by nearby single base insertions (or deletions) several larger deletions are observed when plastid rDNAs are compared with E. coli rDNAs (Schwarz and Kössel, 1980). One of them, a 23 base pair deletion at positions 453-475 appears to be universal for the 5'-terminal domains of chloroplast and cyanobacterial 16S rDNAs (and rRNA, see Fig. 4). 23S rDNA from maize shows insertions of 25, 65 and 78 base pairs when compared with E. coli 23S rDNA (Edwards and Kössel, 1981). The 65 bp insertion seems to be specific for monocotyledons as it is also found in rice (Hiratsuka et al., 1989) but not in the tobacco and liverwort 23S rDNA (Shinozaki et al., 1986; Ohyama et al., 1986). The 78 bp insertion is part of the intergenic region which separates the mature 23S rDNA coding region from the 4.5S rRNA coding region; the insert obviously encodes the signal sequences necessary for cleaving the intergenic region between mature 23S rRNA and 4.5S rRNA from the primary transcript (Strittmatter and Kössel, 1984).

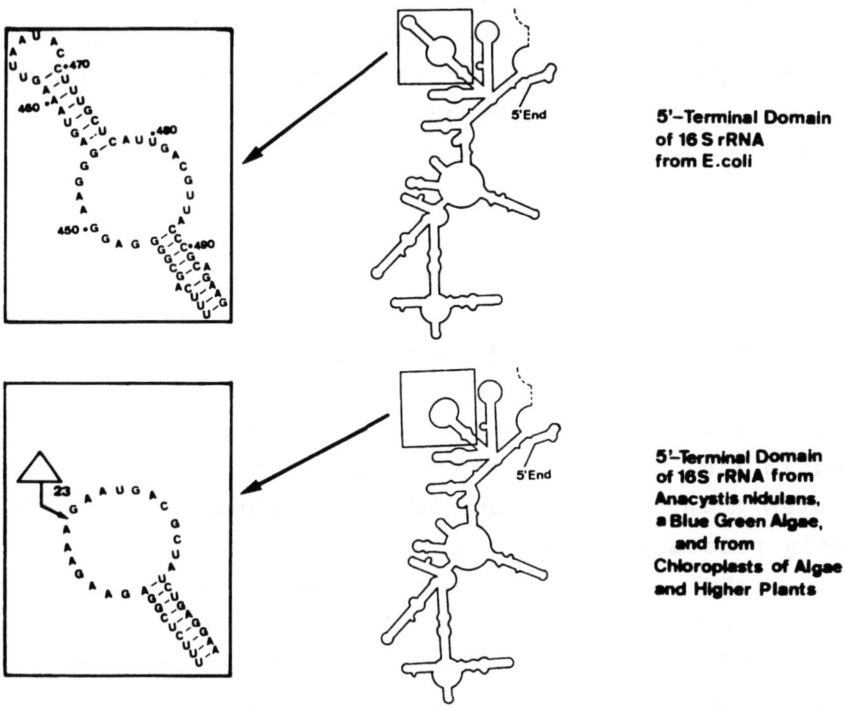

Figure 4. A 23 nt deletion characteristic for cyanobacteria and chloroplast 16S rDNA leads to the amputation of a stem-loop structure in the 5'-terminal domain of the 16S rRNA secondary structure.

Due to compensating base substitutions within base paired regions rRNA <u>secondary structures</u> are in general more highly conservd as compared to the underlying primary structures. In a few cases entire stem loop structures appear either deleted or newly introduced when the secondary structures of plastid rRNAs from various plants and algae are compared with each other or with their bacterial counterparts. For instance, the deletion of 23 bp from maize chloroplast 16S rDNA "amputates" the secondary structure by one complete stem loop structure of the 5'-

terminal domain (Fig. 4). In 23S rDNA from maize chloroplasts the two insertions of 25 and 65 bp cause insertions of extra stem loop structures in the corresponding 23S rRNA secondary structure (Gutell and Fox, 1988). An extra stem loop structure of maize chloroplast 23S rRNA, which is only a "bulge" in the corresponding E. coli 23S rRNA secondary structure, gives rise to one of the prominent hidden breaks (Kössel et al., 1985). It appears feasible that many of the "branches" and "leaves" of the secondary structure "trees" can serve as taxonomic markers in analogous ways to leaves and stems of plants at the macroscopic level.

B. Expression of Chloroplast rRNA Operons

The major steps of chloroplast rRNA operon expression are summarized in Fig. 5 for the rRNA operons of higher plants. As a first step a large precursor rRNA is synthesized which is subsequently processed by several independent steps into the various mature rRNA and tRNA species. Folding into secondary structures is finally followed by assembly with ribosomal proteins into 30S and 50S ribosomal subunits. This expression pathway grossly resembles the well studied bacterial pathway. For this reason the existence of similar enzymes and RNA-substrate structures as for expression of E. coli rRNA operons were anticipated for the chloroplast systems. As deviations from the prokaryotic scenario the splicing activities necessary for spacer tRNAs in higher plants and for 23S rRNA in certain algae (see Fig. 2) and endonuclease activities for the cleavage of small rRNAs (4.5S, 3S and 7S) from the 23S rRNA termini have to be postulated. In contrast to these theoretical considerations, very little detailed information about the various steps has been obtained experimentally.

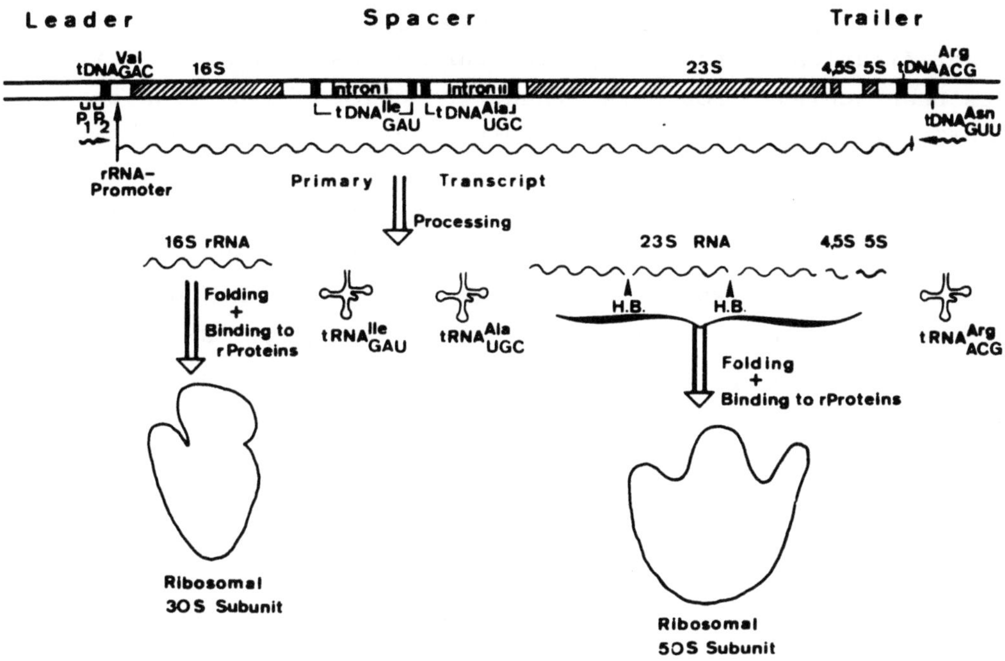

Figure 5. Summary of the various reactions necessary for the expression of the chloroplast rRNA operon of higher plants. H.B., "hidden break".

1. Chloroplast RNA polymeras(es)

The question whether a separate RNA polymease transcribes chloroplast rRNA has been controversial. Originally a complex consisting of chloroplast DNA and tightly bound RNA polymerase termed transctiptional active chromosome (= TAC) was proposed to transcribe specifically rRNA genes (for review see Briat et al., 1986). More recently, however, soluble RNA polymerase from higher plant chloroplasts were shown to transcribe all three classes of RNAs (mRNAs, tRNAs and rRNAs). It is therefore conceivable that TAC is basically a soluble RNA polymerase involved in rRNA synthesis and that its tight linkage to the template rDNA is caused by specific factors or a modified subunit structure of the core enzyme. This view gains support from the rRNA promoter structures which closely resemble promoter structures of mRNA and tRNA encoding genes (see following section).

2. Transcription start site and rRNA promoter structure

Identification of transcription start sites by various criteria has been achieved for more than 10 chloroplast rRNA operons mostly from higher plants (Strittmatter et al., 1985; Delp et al., 1987; Sun et al., 1989) but also from several algae (for review see Delp and Kössel, 1991). The start sites are located 50-200 positions upstream of the respective 16S rRNA genes. Therefore, the $tRNA^{Val}$ sequences encoded further upstream in the rRNA operons of higher plants (see Fig. 2) cannot be included in the primary transcripts of these start sites. In all cases the start site are preceded by -10 and -35 consensus sequences (see Table I) which are typical for prokaryotic promoter structures and which show only minor deviations from plastid mRNA and tRNA promoter structures.

Table I

Promoter consensus sequences:	- 35		- 10
Plastid rRNA Promoters	TTGACRT	—— 14-17 bp ——	TATaYT
Plastid Promoters	TTGaNaN	—— 12-23 bp ——	TAtaaT
E. coli Promoters	TTGACat	—— 15-21 bp ——	TAtAaT

R = Purines; Y = Pyrimidines; N = any of the four bases.

As already mentioned this argues strongly for the existence of only a single core RNA polymerase, which transcribes all RNA classes including rRNAs.

The situation appears to be more complicated in the spinach rRNA operon where two promoters appear to exist in the $tRNA^{Val}$/16S rDNA intergenic region (Briat et al., 1987). Both these promoters are active in the heterologous E. coli system, but only one promoter is active in the homologous spinach system. It remains to be seen whether this unexpected observation of a silent promoter reflects a regulated promoter which is only active under certain physiological conditions.

3. Regulation of plastid rRNA operons

A regulation of plastid rRNA operons by translation mediated attenuation has been suggested based on the existence of several conserved hair pin structures and short open reading frames between the promoter and the 16S rRNA coding leader region. Data obtained in a heterologous E. coli test system lend support to this proposal (Laboure et al., 1988). Final proof, however, requires a homologous test system.

Measurement of steady state level of rRNA and of transcriptional rate levels in vivo (Mullet and Klein, 1987; Klein and Mullet, 1990) contradict the suggestion that plastid rRNA operons are constitutively expressed and regulated only by gene dosage i.e. the multiplicity of chloroplast DNA (Bendich, 1987). A 10-fold decrease of rRNA transcription is observed in 4.5 days old dark grown barley plastids both after continuation of dark and exposure to light (Mullet and Klein, 1987). After growth for 8 days in the dark the transcriptional rate of 16S rRNA (together with mRNAs encoded by the genes rbcL and psbA) can, however, be reinduced up to 5-fold (ca. 50% of the level after 4.5 days in the dark) by exposure to light (Klein and Mullet, 1990). Similar observations were obtained with 16S rRNA transcription in maize plastids by illumination of seedlings grown for 9 days in the dark. This clearly demonstrates that plastid rRNA operons are subject to regulation in dependence of developmental stage and/or light. This is not in contradiction to parallel observations on the steady state levels of 16S rRNA, which are less sensitive or even independent of the developmental stage and/or light induction indicating that steady state levels are regulated largely by the balance between stability and breakdown and only in part by transcription.

No regulatory cis acting regulatory elements (besides the above mentioned proposal of an attenuator) have been identified in the 16S rRNA leader region but the possibility exists that the topoisomeric state of this region regulates the transcriptional rate. This has been demonstrated in an in vitro system derived from pea chloroplasts, in which the 16S promoter is most active in the negatively supercoiled form (Sun et al., 1989).

4. Termination of the rRNA operons

In spite of the failure to detect 5S rRNA sequences in the large rRNA precursor, cotranscription of 5S rRNA could be demonstrated by S_1 and primer extension techniques in maize (Strittmatter and Kössel, 1984) and spinach (Audren et al., 1987). Therefore an independent 5S rRNA transcription appears unlikely and the failure to detect 5S rRNA sequences in the primary transcript is probably due to rapid cleavage of 5S rRNA from the precursor rRNA.

The question whether $tRNA^{Arg}$ is also included or whether termination already occurs in the 5S rDNA/$tDNA^{Arg}$ intergenic region (see Fig. 5) and $tRNA^{Arg}$ represents a separate transcriptional unit remains unsolved. In view of the existence of most of the intergenic RNA (as shown in maize by S_1 and primer extension technique; Dormann-Przybyl et al., 1986) and because of functional homology with the cotranscribed trailer tRNA genes of bacterial rRNA operons (see Fig. 2), the possibility of a cotranscription of $tRNA^{Arg}$ and termination of the primary rRNA transcript distal to this trailer tRNA gene appears more likely. Development of functional test systems for transcriptional termination of plastid genes will be essential to resolve this question.

5. Processing of the primary rRNA transcript

Dissection of the primary rRNA transcript to mature rRNAs and tRNAs requires several processing enzymes such as ribonucleases for endonucleolytic cleavage and exonucleolytic trimming of rRNA and tRNA flanking sequences, splicing activities for spacer tRNAs or 23S rRNA transcripts containing intervening sequences, CCA terminal transferase for tRNA maturation and enzymes for base modifications in both, rRNAs and tRNAs. With the exception of tRNA specific RNases, in particular RNase P (Wang et al., 1988), which act on the spacer tRNA sequences of the primary rRNA transcript, no processing enzymes specifically involved in rRNA processing and maturation have so far been characterized from plastids. However, putative processing sites have been identified by S_1 and reverse transcriptase mapping (Strittmatter and Kössel, 1984; Dormann-Przybyl et al., 1986). The folding of 5'- and 3'-terminal precursor sequences of 16S rRNA (Strittmatter et al., 1985) and of the 23S/4.5S rRNAs (Edwards and Kössel, 1981) to double stranded stem structures similar to the respective E. coli rRNA precursors is per se a strong argument for the involvement of an <u>RNase III like activity</u> in endonucleolytic cleavage of plastid rRNA primary transcripts. The identification of a limited number of modified bases in plastid 16S rRNAs especially of N^6-dimethyladenosine within the 3'-terminal loop (for review see Delp and Kössel, 1991) clearly shows the existence of base modifying enzymes. But further investigations are necessary to identify other modified rRNA positions and to characterize the respective enzymes.

The occurrence of <u>hidden breaks</u> (H.B., see Fig. 5) is specific for the 23S rRNA of higher plant chloroplasts (for review see Delp and Kössel, 1991). In view of the specificity and their dependence on leaf development it is likely that hidden breaks reflect processing steps necessary for certain unknown structural requirements and/or functions of plastid ribosomes. The positions of the two major hidden breaks have been determined in maize chloroplast 23S rRNA by S_1 mapping (Kössel et al., 1985). They coincide with stem loop structures which in comparison with bacterial 23S rRNA are either specifically inserted or reduced in size. Extra stem loop structures per se which are present in other positions of the 23S rRNA secondary structure (see above) do not lead to hidden breaks. It is, therefore, not clear which additional structural features or nucleases are causing the specific cleavage reactions leading to the hidden breaks.

Attempts in several laboratories to demonstrate and characterize <u>splicing activities</u> from chloroplasts have met with very limited success so far. Splicing of a nonradioactive spacer

$tRNA^{Ala}$ precursor (with 5' and 3' extensions and the 806 nt class II intron) synthesized in vitro appears to be possible by using a tRNA free chloroplast extract from maize as analysed by postlabelling of the tRNA sized product by CCA addition (Delp, Igloi and Kössel, unpublished). In addition to this the polymerase chain reaction technique has been used for amplification and sequencing of spacer tRNA processing intermediates from total maize chloroplast RNA (Delp and Kössel, unpublished). This approach has permitted the exact localization of the splice junction of mature and 5' extended tRNA. The failure to amplify 3' extended and/or unspliced tRNA precursors indicates that 5' maturation - contrary to earlier suggestions (Wang et al., 1988) - may occur as the last step after 3' maturation and splicing.

C. Conformational switch of small ribosomal subunit RNAs

The potential base pairing of 5'- and 3'-flanking sequences of 16S rRNA (Strittmatter et al., 1985) mentioned above together with 5'/3'-terminal pairing in many mature RNAs such as 23S rRNAs, 5S rRNAs, tRNAs and M1 RNA has led us to perform a screening for 5'/3'-complementary sequences in all known 16S rRNA sequences. As a result highly conserved sequences were found which allow 5 or 6 base pairs between the two terminal regions of bacterial, plastid, archaebacterial and eukaryotic small subunit rRNAs. Formation of the 5'/3' helical arrangement is, however, only possible at the cost of dissolving the pseudo-knot helix between the 5'-terminal region and the internal region of small subunit rRNA. It is postulated (Kössel et al., 1990) that the mutually exclusive conformational states are in dynamic equilibrium (Fig. 6), in which the pseudo-knot containing conformation with the 3'-terminal sequences more exposed is likely to represent the initiating state of the samll ribosomal subunit, whereas the 5'/3'-terminally paired "closed" conformation may represent the elongating state in which interaction with fortuitous ribosomal binding sites of mRNAs is avoided.

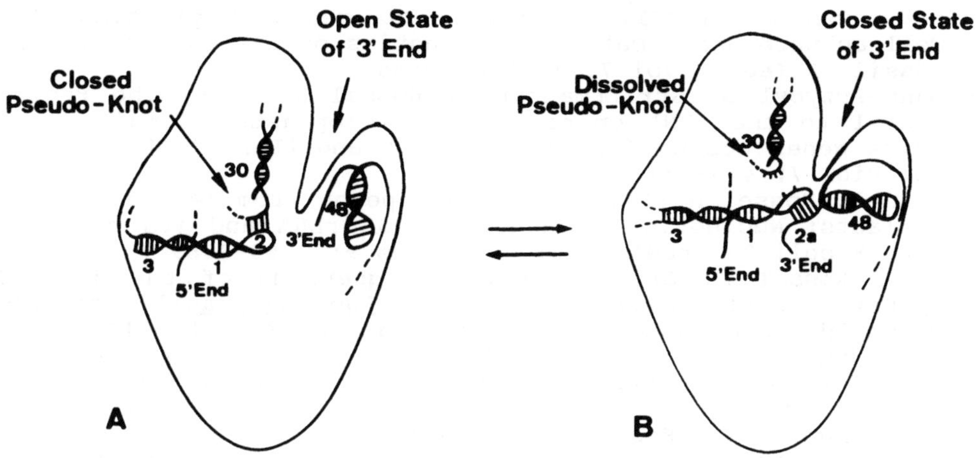

Figure 6. A. Position and orientation of helices 1, 2, 3, 30 and 48 in the three-dimensional structure of the E. coli 30S subunit. B. Position and orientation of the same element after the conformational switch to the closed state of the 3'-terminus. In both cases the 30S/50S interface side of the subunit is depicted (according to Kössel et al., 1990).

Acknowledgements: The continuous support of this work by several grants from the Deutsche Forschungsgemeinschaft (SFB 46 and 206), the Bundesministerium für Forschung und Technologie and the Fonds der Chemischen Industrie is gratefully acknowledged.

References

Audren H, Bisanz-Seyer C, Briat J-F, Mache R (1987) Structure and transcription of the 5S rRNA gene from spinach chloroplasts. Curr Genet 12: 263-270

Bendich AJ (1987) Why do chloroplast and mitochondria contain so many copies of their genome? BioEssays 6: 279-282

Briat JF, Lescure AM, Mache R (1986) Transcription of the chloroplast DNA: a review. Biochimie 68: 981-990

Briat JF, Bisanz-Seyer C, Lescure A-M (1987) In vitro transcription initiation of the rDNA operon of spinach chloroplast by a highly purified soluble homologous RNA polymerase. Curr Genet 11: 259-263

Delp G, Igloi GL, Beck F, Kössel H (1987) Functional in vivo verification in E coli of promoter activities from the rDNA/tDNA^{Val}(GAC) leader region of Zea mays chloroplasts. Curr Genet 12: 241-246

Delp G, Kössel H (1991) rRNAs and rRNA genes of plastids. In: Cell culture and somatic cell genetics of plants. Bogorad L, Vassil IK (eds). Vol 7, in the press

Dormann-Przybyl D, Strittmatter G, Kössel H (1986) The region distal to the rRNA operon from chloroplasts of maize contains genes coding for $tRNA^{Arg}$(ACG) and $tRNA^{Asn}$(GUU). Plant Mol Biol 7: 419-431

Edwards K, Kössel H (1981) The rRNA operon from Zea mays chloroplasts: Nucleotide sequence of 23S rDNA and its homology with E coli 23S rDNA. Nucl Acids Res 9: 2853-2869

Graf L, Kössel H, Stutz E (1980) Sequencing of the 16S-23S spacer in a ribosomal RNA operon of Euglena gracilis chloroplast DNA reveals two tRNA genes. Nature 286: 908-910

Graf L, Roux E, Stutz E, Kössel H (1982) Nucleotides sequence of a Euglena gracilis chloroplast gene coding for the 16S rRNA: Homologies to E. coli and Zea mays chloroplast 16S rRNA. Nucl Acids Res 10: 6369-6381

Gruissem W (1989) Chloroplast RNA: Transcription and processing. In: Biochemistry of Plants. Marcus A, (ed). Vol. 15: 151-191 Academic Press

Gutell RR, Fox GE (1988) A compilation of large subunit RNA sequences presented in a structural format. Nucl Acids Res 16: r175-r269

Hiratsuka J, Shimada H, Whittier R, Ishibashi T, Sakamoto M, Mori M, Kondo C, Honji Y, Sun C-R, Meng B-Y, Li Y-Q, Kanno A, Nishizawa Y, Hirai A, Shinozaki K, Sugiura M (1989) The complete sequence of the rice (Oryza sativa) chloroplast genome: Intermolecular recombination between distinct rRNA genes accounts for a major plastid DNA inversion during the evolution of the cereals. Mol Gen Genet 217: 185-194

Klein RR, Mullet JE (1990) Light-induced transcription of chloroplast genes. J Biol Chem 265: 1895-1902

Kössel H, Natt E, Strittmatter G, Fritzsche E, Gozdzicka-Jozefiak A, Przybyl D (1985) Structure and expression of rRNA operons from plastids of higher plants. In: Molecular form and function of the plant genome. Vloten-Doting L van, Groot GSP, Hall TC (eds). Plenum Publishing Corporation, pp 183-198

Kössel H, Hoch B, Zeltz P (1990) Alternative base pairing between 5'- and 3'-terminal sequences of small subunit RNA may provide the basis of a conformational switch of the small ribosomal subunit. Nucl Acids Res 18: 4083-4088

Koch W, Edwards K, Kössel H (1981) Sequencing of the 16S-23S spacer in a ribosomal RNA operon of Zea mays chloroplast DNA reveals two split tRNA genes. Cell 25: 203-213

Laboure A-M, Lescure A-M, Briat J-F (1988) Evidence for a translation-mediated attenuation of a spinach chloroplast rDNA operon. Biochimie 70: 1343-1352

Lyttelton JW (1960) Nucleoproteins of white clover. Biochem J 74: 82-90

Manhart JR, Palmer JD (1990) The gain of two chloroplast tRNA introns marks the green algal ancestors of land plants. Nature 345: 268-270

Mullet JE, Klein RR (1987) Transcription and RNA stability are important determinants of higher plant chloroplast RNA levels. EMBO J 6: 1571-1579

Ohyama K, Fukuzawa H, Kohchi T, Shirai H, Sano S, Umesono K, Shiki Y, Takeuchi M, Chang Z, Aota S, Inokuchi H, Ozeki H (1986) Chloroplast gene organization deduced from complete sequence of liverwort Marchantia polymorpha chloroplast DNA. Nature 322: 572-574

Pace NR, Olsen GJ, Woese CR (1986) Ribosomal RNA phylogeny and the primary lines of evolutionary descent. Cell 45: 325-326

Palmer JD (1985) Comparative organization of chloroplast genomes. Ann Rev Genet 19: 325-354

Przybyl D, Fritzsche E, Edwards K, Kössel H, Falk B, Thompson JA, Link G (1984) The ribosomal RNA genes from mustard (Sinapis alba L.): Mapping and sequencing of the leader region. Plant Molec Biol 3: 147-158

Rochaix J-D (1987) Molecular genetics of chloroplast and mitochondria in the unicellular green alga Chlamydomonas. FEMS Microbiology Rev 46: 13-34

Ruf M, Kössel H (1988) Occurence and spacing of ribosome recognition sites in mRNAs of chloroplasts from higher plants. FEBS Lett 240: 41-44

Schwarz Zs, Kössel H (1980) The primary structure of 16S rDNA from Zea mays chloroplast is homologous to E coli 16S rRNA. Nature 283: 739-742

Schwarz Zs, Kössel H, Schwarz E, Bogorad L (1981) A gene coding for $tRNA^{Val}$ is located near the 5' terminus of the 16S rRNA gene in the Zea mays chloroplast genome. Proc Natl Acad Sci USA 78: 4748-4752

Shinozaki K, Ohme M, Tanaka M, Wakasugi T, Hayashida N, Matsubayashi T, Zaita N, Chunwongse J, Obokata J, Yamaguchi-Shinozaki K, Ohto C, Torazawa K, Meng BY, Sugita M, Deno H, Kamogashira T, Yamada K, Kusuda J, Takaiwa F, Kato A, Tohdoh N, Shimada H, Sugiura M (1986) The complete nucleotide sequence of the tobacco chloroplast genome: its gene organization and expression. EMBO J 5: 2043-2049

Steinmetz A, Weil J-H (1989) Protein synthesis in chloroplasts. In: Biochemistry of Plants Marcus A, (ed). Vol. 15: 193-227. Academic Press

Strittmatter G, Kössel H (1984) Cotranscription and processing of 23S, 4.5S and 5S rRNA in chloroplasts from Zea mays. Nucl Acids Res 12: 7633-7647

Strittmatter G, Gozdzicka-Jozefiak A, Kössel H (1985) Identification of an rRNA operon promoter from Zea mays chloroplasts which excludes the proximal tRNAVal (GAC) from the primary transcript. EMBO J 4: 599-604

Sugiura M (1989) The chloroplast chromosomes in land plants. Annu Rev Cell Biol 5: 51-70

Sun E, Wu B-W, Tewari KK (1989) In vitro analysis of the pea chloroplast 16S rRNA gene promoter. Mol Cell Biol 9:5650-5659

Wang MJ, Davis NW, Gegenheimer P (1988) Novel mechanisms for maturation of chloroplast transfer RNA precursors. EMBO J 7: 1567-1574

Yamada T, Shimaji M (1987) Splitting of the ribosomal RNA operon on chloroplast DNA from Chlorella ellipsoidea. Mol Gen Genet 208: 377-383

RIBOSOMAL RNA GENES AND PSEUDOGENES OF THE BI-MOLECULAR PLASTID GENOME OF THE BROWN ALGA *Pylaiella littoralis*.

S. Loiseaux-de Goër, Y. Markowicz*, and C.C. Somerville
Station Biologique de Roscoff, CNRS-UPR4601
B.P. 74
29680 Roscoff Cedex, France

Introduction

The algae have long been classified into three main groups based on their plastidial pigments (Christensen, 1964). In this scheme all three groups have the same primary pigment, chlorophyll *a*, and the different taxa are defined by their secondary pigments; the green algae having chlorophyll *b*, the red algae having phycobilins, and the chromophytes (which include the brown algae), having chlorophyll *c*. Today, plastidial characters are still viewed as important in defining the relationships among the algae, and between the algae and higher plants. In this light, some of the most intriguing questions about the chromophytes concern the different morphologies, origins, and evolutionary histories of their plastids. One way to study these organelles is to examine them at the genomic level and to compare both the gross organization and fine structure of their genes with those of other plastids (not only green plastids) and prokaryotes (not only cyanobacteria). For this purpose, we have begun an extensive investigation of the plastidial genome of a simple brown alga, *Pylaiella littoralis* (L.) Kjellm. Herein we describe progress to date in characterizing the ribosomal operons, genes and pseudogenes found on this genome, and we discuss the phylogenetic implications of these results.

Structure of the ribosomal operons of *P. littoralis*

In *P. littoralis* the genome of the chloroplast is composed of two different circular DNA molecules; one of 133 kbp and one of 58 kbp

*Laboratoire Biologie Moleculaire Végétale, CNRS-URA 57, Université Joseph Fourier 1, , B.P. 53X, 38041 Grenoble Cedex, France

NATO ASI Series, Vol. H 55
The Translational Apparatus of Photosynthetic Organelles
Edited by R. Mache et al.

(Loiseaux-de Goër, et al., 1988). This type of bimolecular genome is relatively rare, having been found only in the plastids of some members of the Chromophyta (Cattolico and Loiseaux-de Goër, 1989). As in most plants, the plastidial genome carries two complete copies of the ribosomal RNA operon on inverse repeats. In *Pylaiella,* they occur on the larger (133 kbp) of the two genomic molecules and are separated by 45 kbp (see Figure 1, Loiseaux-de Goër, et al. 1988). As in *Dictyota* (Kuhsel and Kowallik, 1987), the inverse repeats are smaller than those of green plants (< 5 kbp), and apparently contain only rRNA and tRNA genes. In addition, the smaller DNA molecule (58 kbp) carries both a 16S rRNA pseudogene with only 65% sequence homology to the functional 16S rRNA gene, and a region which hybridizes with the 23S rRNA. The 23S-hybridizing region is quite complex; it is composed of two sequences separated by a 3 kbp region which does not hybridize to the probe. It may represent a split gene, as found in *Chlorophyceae*, or a long-inactive pseudogene which is conserved enough to give hybridization signals (Loiseaux-de Goër, et al., 1988). This region of the small molecule has been cloned, and is currently being sequenced. The 16S rRNA pseudogene will be discussed in some detail in a subsequent section.

The ribosomal RNA operons of *P. littoralis*, which contain genes for the 16S rRNA, $tRNA^{Ala}$, $tRNA^{Ile}$, the 23S rRNA, and the 5S rRNA, are relatively short due to the reduced lengths of the intergenic non-coding regions. For example, the spacer region between the two tRNA genes is only 3 nucleotides in length (Markowicz et al., 1988a), whereas the corresponding regions in prokaryotes and higher plants contain an average of 30 and 80 nucleotides, respectively. Also, the spacer region between the 23S and the 5S rRNA genes in *P. littoralis* contains only 25 nucleotides compared to more than 200 nucleotides in higher plants.

A comparison among known ribosomal operons (Figure 1) shows two distinct groups. The *Pylaiella* operons are clearly more similar to those of prokaryotes, *Euglena*, and the cyanelle of *Cyanophora paradoxa* than they are to the *rrn* operons of green algae and land plants, *i.e.*, the "green lineage". In the "green lineage" the spacer regions between the 16S and 23S rRNA genes are either very enlarged or are completely disorganized, as in the case of *Chlorella ellipsoidea* (*Chlorophyceae*), where the inversion of a 5 kbp DNA fragment leads to a unique organization of the operon (Yamada and Shimajii, 1986). In extreme cases, such as the green alga *Codium fragile* (*Ulvophyceae*; not shown) where the 16S and 23S rRNA genes are separated by 8 kbp, the ribosomal operon seems to be completely lost (although the transcription of these genes has not been characterized; Manhart et al., 1989).

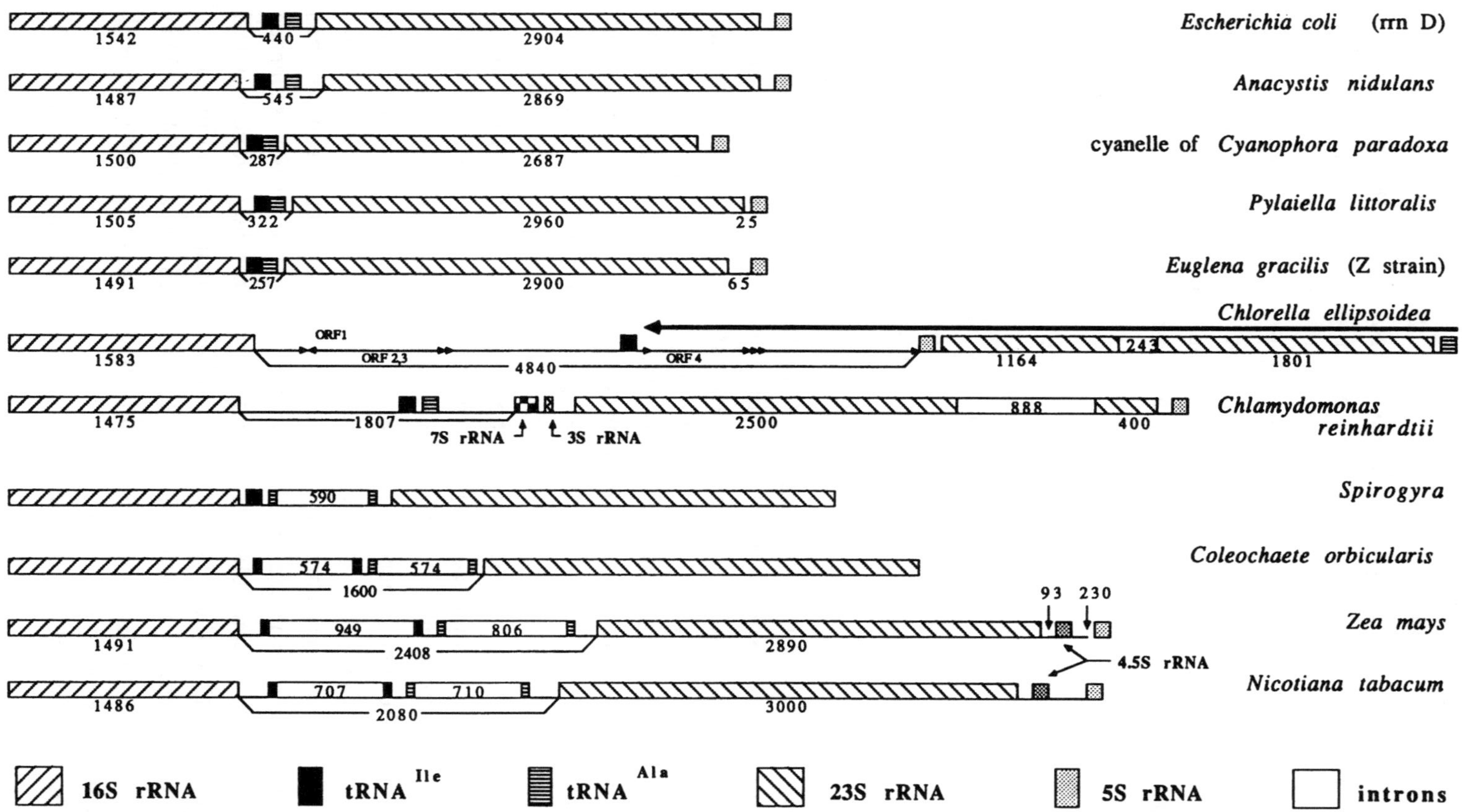

Figure 1. Comparison of prokaryotic, cyanellar, and plastidial rRNA operons. The gene order and position of intergenic sequences and introns are shown for the organisms listed at the right. The relative sizes of the operons are indicated for gross comparison, but the sizes of individual genes or spacers are not strictly to scale. The exact sizes of genes and intergenic sequences are indicated below each map (where available), and intron sizes are indicated within the open boxes.

Another characteristic feature of the *rrn* operons of the "green lineage" (which, by this definition, does not include *Euglena*) is the presence of introns. Class II introns are present in both $tRNA^{Ile}$ and $tRNA^{Ala}$ genes of all land plants and *Charophyceae* with the exception of *Spirogyra*, a more primitive Charophyceaen, where only the $tRNA^{Ala}$ gene contains an intron (Manhart and Palmer, 1990). This intermediary condition in terms of introns suggests that the land plants emerged from the same lineage as the *Charophyceae*. In another class of green algae, the *Chlorophyceae*, introns have been found in the 23S rRNA gene, and at least one has been shown to be a mobile group I intron which contains the gene for an endonuclease (Lemieux et al., 1988).

Manhart and Palmer (1990) argue that the presence of introns in these operons is a recent evolutionary occurrence. If it is assumed that all plastids, and hence all the plastidial *rrn* operons shown in Figure 1, have ultimately arisen from an endosymbiotic relationship between a heterotrophic (organizationally eukaryotic) host and a prokaryotic phototroph, then the ancestral *rrn* operon was probably free of introns, which have since invaded the green plastid line. Note that such an assumption does not rule out a secondary eukaryote-eukaryote symbiosis giving rise to some plastidial lines. This conclusion is much more parsimonius than the opposing assumption that introns were present in the ancestor, but have been subsequently lost in all prokaryotic and plastidial lineages except that of the green plastids (Manhart and Palmer, 1990). Alternately, it is possible to explain the differences in *rrn* organization by assuming a polyphyletic history for plastids, and thus avoiding the limitations imposed by parsimony. And, in fact, there is both morphological and molecular evidence to suggest that all contemporary plastids can not be explained by a single cyanobacterial-like plastid ancestor (Gibbs, 1981; Markowicz et al., 1988b; Gibbs, 1990; Assali et al., 1990). In this case, one might assume that the *rrn* operons of the "green lineage" evolved from another progenitor, perhaps one which carried introns. The data, however, do not support the latter assumption. Gene sequences from green plastids are consistent with a cyanobacterial-like ancestor, and 16S rRNA sequence data not only place green plastids directly within the cyanobacterial lineage, but also suggest that cyanobacterial diversification occurred much later than that of the major eubacterial lines (Giovannoni, et al.,1988). Thus it is very difficult to imagine any scenario in which *rrn* operons with introns predated those without introns.

Again, the question of interest to us is the evolutionary history of the brown plastids, since these have properties suggesting a phototrophic eukaryotic endosymbiont (Gibbs, 1981; Gibbs, 1990), but the organization of their *rrn* operons is clearly more similar to those of contemporary prokaryotes. In order to gain further insight about the phylogeny of these operons, we analysed the sequences of the 16S rDNA.

The 16S rRNA gene and pseudogene of *P. littoralis*

The 16S rDNA of the ribosomal RNA operon and the 16S pseudogene found on the small DNA molecule of *P. littoralis* have been sequenced, and their secondary structures established (Markowicz et al., 1988b). The 16S rRNA gene of *P. littoralis* is a typical plastidial gene, showing well conserved prokaryotic features (71 out of 72 eubacterial signatures; Woese, 1987), and also one of the four specific beta-purple bacterial signatures. The beta-purple bacteria, as defined by Woese (1987), include photosynthetic procaryotes as well as common gram-negative bacteria such as *Alcaligenes* and some Pseudomonads, but do not include the cyanobacteria. The functional gene and the pseudogene share specific base sequences not found in *Escherichia coli*, cyanobacteria, *Euglena*, or green plants. Unfortunately, because of the lack of other data, no comparison has been made with genes of other Chromophytes or those of Rhodophytes. Therefore, one can not discriminate between the duplication of the gene in *P. littoralis* followed by inactivation and sequence drift, and different (although closely related) origins for these sequences which are carried on separate DNA molecules. In either case, the genes are now only 65% homologous, indicating that they have evolved separately for a relatively long time. The mutations found in the helical regions of the pseudogene have occurred in a complementary manor such that the secondary structure is still well conserved. This suggests that the inactivation of the pseudogene occurred more recently than its divergence from the active gene. Sequence drift between the two genes has been greatest in the 3' minor domain, with only 45% homology remaining in this area. The pseudogene has lost the GGAA (nucleotide numbers 1478-1482) loop, mutated in the Shine-Dalgarno sequence, and does not have the antitermination-like sequence found upstream of the active gene (Figure 2). This supports the finding that this gene is not expressed during any phase of cell growth (Markowicz et al., 1988b). It is not known, however; which, if any, of these changes caused inactivation of the gene, and which have arisen after loss of expression.

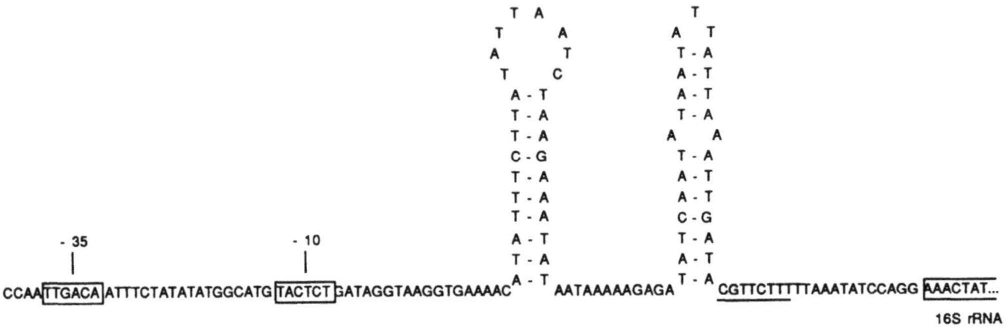

Figure 2. The antitermination-like sequence found upstream of the active 16S rRNA gene of *P. littoralis* (Markowicz,1988. Dissertation, Université Joseph Fourier, Grenoble 1, France)

A phylogenetic tree inferred from comparison of 16S rRNA gene sequences shows that the plastids of *P. littoralis* and *Euglena* share a recent common ancestor, and that the *Pylaiella-Euglena* line diverged from the original cyanobacterial lineage slightly later than the line which gave rise to plastids of green algae and land plants (Markowicz, et al., 1988b). Since endosymbioses must have occured after the divergence from the lineage which includes the extant cyanobacteria, the establishment of plastids in both the *Euglena-Pylaiella* and higher plant lines requires more than one plastidial ancestor established in separate symbioses. This agrees well with the fact that phylogenetic trees constructed from sequences of the cytoplasmic (or eukaryotic) ribosomal RNAs from *Euglena, Chromophyta*, and members of the "green lineage" show that they represent distinct and independent lineages (Gunderson, et al., 1987; Perasso, et al., 1989). This is particularly obvious in the case of *Euglena*, which diverged very early from the remainder of the eukaryotic tree, but acquired plastids as late as other plants, implying independent establishment of the phototrophic endosymbiont.

Organization of the 3' end of the ribosomal operon of *P. littoralis*

The 3' end of the 23S rRNA gene of *P. littoralis*, which shares significant homology with the 4.5S rRNA species of higher plants, has been sequenced. The similarity of this region to those of prokaryotic and plastidial sequences and the absence of an identifiable spacer region suggest that the independent 4.5S rRNA species does not exist in *Pylaiella* as it does in higher plants. As mentioned above, however, the spacer regions of the *Pylaiella rrn* operon are much reduced in comparison to both prokaryotic and eukaryotic *rrn* operons, and, therefore, the status of this region must be confirmed by further rRNA characterization. As shown in Figure 3, the sequence has been aligned with equivalent sequences from eubacteria, cyanobacteria, *Euglena*, and the 4.5S rRNA sequences of green plastids. The *Pylaiella* sequence shares 57.3% homology with *E. coli* , 64.5% with *Euglena,* 67.0% with *Synechococcus* (*Anacystis*) *nidulans*, and an average of 72.3% with higher plants (sequences aligned by eye, all homologies based on the number of identical bases in the aligned sequences divided by the true length of the shorter sequence in each pair). The extent of the sequence homology with higher plants is somewhat suprising, given the relationships delineated by the 16S rRNA sequence (*vide supra*), and may be misleading due to the presence of "insert" regions in the *Pylaiella* sequence and in those of the green mosses, ferns, and dicotyledons (Figure 3) which interupt the alignment of the genes. Parsimony would argue that these regions represent insertions in the ancestral sequence since they occur within a closely related group (with the exception of *Pylaiella*, *vide infra*), and are absent among those sequences representing a much deeper phylogenetic assemblage. The *Pylaiella* inserts are unique both in their number (two inserts in *Pylaiella* versus a maximum of one in all other sequences) and sizes (19 and 12 nucleotides in length as opposed to a maximum of 9 nucleotides in the other sequences). The 19 nt insert is found in the same locus as the 9 nt insertion common to green mosses, ferns and dicotyledons, and occurs within a highly conserved region. Insertions of one or two nucleotides are found at the same locus in monocotyledons and in *Chlorella* and *Euglena*. The second insertion, 12 nucleotides long, is apparently unique to *P. littoralis*, and changes the secondary structure of the 23S rRNA by increasing the length of the 3'-terminal helix (not shown).

The 3' end of the 23S rRNA is separated by a spacer of 25 nucleotides from the putative 5S rRNA gene (Figure 4). This spacer is the shortest described to date. Other examples include the 65-nucleotide spacer in *Euglena* which has no homolgy with the *Pylaiella* sequence, and the 222- to 261-nucleotide spacers found in higher plants (Strittmatter and Kossel, 1984; Audren, et al., 1987; Raafar El-Gewely, et al., 1984). The *Pylaiella* spacer sequence contains no direct or inverse repeats, and no promoter-like sequences, both of which are found in this spacer region in higher plants.

```
E. coli        23S   AAGGA-ACGTTGAAGACGACGACGTT---------GATAGGCCGGGTGTGTAAGCGCAGCGATGCGTTGAGCTAACCGGTACTAAT-GAACCGTGA-GG-CTTAACCTT
P. aeruginosa  23S   AAGGG-CCGTCGAAGACTACGACGTT---------GATAGGCTGGGTGTGTAAGCGTTGTGAGGCGTTGAGCTAACCAGTACTAAT-TGCCCGTGA-GG-CTTGACCAT
B. subtilis    23S   AAGAT-CCCTGAAAGATGATCAGGTT---------GATAGGTCTGAGGTGGAAGTGTGGCAACACATGGAGCTGACAGATACTAAT-CGATC--GA-GGACTTAACCAT
B. stearo.     23S   AAGAT-CCCTCGAAGATGACGAGGTC---------GATAGGTCCGAGGTGGAAGCGTGGCGACACGTGGAGCTGACGGATACTAAT-CGATC--GA-GGGCTTAACCTA
A. nidulans    23S   AAGGT-CACGGGTAGAACACC-CGTT---------AATAGGCGCTATGTGGAAGTTCAGCAATGGATGAAGCTGAGGCGTACTAAT-AGACC--GA-GGGCTTGACCTC
Euglena        23S   -AGGT-CACGGTAAGACGAAC-CGTTT--------GATAGGTATTAGGTGTACAATTGGTAACAATTTTAGCCGAGATATACTAACCGA-CCGAAAATTTTTTCC----
Chlorella      23S   AAGGT-CACAGTAAGACTAAC-TGTTC--------GATAGCTATCAAGTGTAAGGCTAGTAATAGTTGCAGCTGAGATAGACTAACAGA-CCGTAATGAATTTT-----
Corn           4.5S  AAGGT-AGCGGCGAGACGAGC-CGTTTA-------AATAGGTGTCAAGTGGAAGTGCAGTGATGTATGCAGCTGAGGCATCCTAAC-GAA-CGAAC-GA-TTTGAAC--
Wheat          4.5S  AAGGTGAGCGGCGAGACGAGC-CGTTTA-------AATAGGTGTCAAGTGGAAGTGCAGTGATGTATGCAGCTGAGGCATCCTAAC-GAA-CGAAC-GA-TTTGAAC--
Tobacco        4.5S  AAGGT-CACGGCGAGACGAGC-CGTTTATCATTACGATAGGTGTCAAGTGGAAGTGCAGTGATGTATGCAGCTGAGGCATCCTAACAGA-CCGGTA-GA-CTTGAAC--
Spinach        4.5S  AAGGT-CACGGCGAGACGAGC-CGTTTATCATTACGATAGGTGTCAAGTGGAAGTGCAGTGATGTATGCAGCTGAGGCATCCTAACAGA-CCCACA-GA-CTTGAAC--
Liverwort      4.5S  AAGGT-CACGGCAAGACTAGC-CGTTTATTTTTACGATAGGTGCCAAGTGGAAGTGCAGTAATGTATGTAGCTGAGGCATCCTAACAGAACCGAGA-GA-TTTGAAC--
Fern           4.5S  AAGGT-CACGGCAAGACGAGC-CGTTTATCACCACGATAGGTGCTAAGTGGAGGTGCAGTAATGTATGCAGCTGAGGCATCCTAATAGA-CCGAGA-GG-TTTGAAC--
Pylaiella      23S?  AAGAT-CACGGCAAGACTAGC-CGTTCATTTCGGGGATAGTTATCAAGTAGAAGTATAGCAATATATTAAGCTGAGATGTACTAACAGA-TCGA---GGACTTGAACTT
                                                 TTACTATACC    TACGATAGGCCA
```

Figure 3. Alignment of the 4.5S rRNAs and 4.5S-like regions of the 23S rRNAs of bacterial and plastidial genes. Conserved bases are shown in bold type. The insert regions which are discussed in the text are delineated. Abbreviations: E. coli, *Escherichia coli;* P. aeruginosa, *Pseudomonas aeruginaosa* (γ-purple eubacteria); B. subtilis, *Bacillus subtilis* ;B. stearo., *Bacillus stearothermophilus* (gram-positive eubacteria); A. nidulans, *Anacystis nidulans* (cyanobacterium); Euglena, *Euglena gracillis* (photosynthetic protist, plastid); Chlorella, *Chlorella ellipsoidea* (green alga, plastid); Corn, *Zea mays* (plastid); Wheat, *Triticum aestivum* (plastid); Tobacco, *Nicotiana tabacum* (plastid); Spinach, *Spinacea oleracea* (plastid); Liverwort, *Marchantia polymorpha* (plastid); Fern, *Dryopteris acuminata* (plastid); Pylaiella, *Pylaiella littoralis* (brown alga, plastid).

TTGAAC TTTTTTCTAGCTTTAAAAATTATGG TCTTGG

23S rRNA 5S rRNA

Figure 4. Sequence of the spacer region between the 23S rRNA gene and the putative 5S rRNA gene.

The sequence of the putative 5S rRNA gene is generally well conserved, having homologies of 70.3% with the cyanellar 5S rRNA of *Cyanophora paradoxa*, 63.0% with the chloroplast gene of spinach, 62.2% with *Prochloron*, 50.4% with *E. coli*, 46.1% with *Euglena gracilis*, and 40.3% with the cytoplasmic 5S rRNA of *C. paradoxa* (sequences were aligned according to Specht, et al., 1990, and homologies calculated as described above). However, as was the case for the 4.5S-like sequence, the 5S rRNA apparently folds into a secondary structure which is unique to *Pylaiella* (not shown). A thorough study of this gene and its implications in the phylogeny of the chloroplasts of *P. littoralis* is currently in progress.

Conclusions

The gross organization of the plastidial ribosomal RNA operon of the brown alga *P. littoralis*, is similar to that of the *rrn* operons of the eubacteria, cyanobacteria, the cyanelles of *Cyanophora*, and *Euglena*. They differ markedly from the ribosomal RNA operons found in plastids of the "green lineage" which are enlarged, sometimes more or less disorganised, and which carry introns. At the level of the nucleotide sequence, however, the *Pylaiella* operon is distinct from both the "primitive" group and the "green lineage", having much reduced spacer regions between the individual RNA genes, and unique secondary structures in both the 4.5S-like region, and the putative 5S rRNA. Together with the morphological data, these studies suggest that the plastids of chromophytes represent a distinct, and potentially very interesting, branch in the phylogeny of photosynthetic organelles. This contention is supported by recent publications (Boczar, et al., 1989; Assali, et al., 1990) which show that some of the plastidial genes of brown algae are more closely related to the analogous genes of beta-purple eubacteria than to their analogs in cyanobacteria or green plastids. A more definitive view of the interrelationships among *Pylaiella*, the eubacteria, and other algal and plant plastids will be possible upon completion of the entire 23S rRNA gene sequence. The 23S rRNAs are longer and somewhat less conserved than the 16S rRNAs, and should provide better resolution of the order of emergence of the various eubacterial, cyanobacteria, and plastidial lines of descent; all of which appeared during a relatively short evolutionary period. In the same way studies on the 23S rRNA-hybridizing region found on the small DNA molecule of *P. littoralis* may provide more information on the origin and evolution of rRNA genes and pseudogenes in chromophyte plastids.

Bibliography

Assali N.E, Mache R., and Loiseaux-de Goër S. (1990) Evidence for a composite phylogenetic origin of the plastid genome of the brown alga *Pylaiella littoralis* (L.) Kjellm. Plant Mol. Biol. (in press).

Audren H., Bisanz-Seyer C., Briat J.F., and Mache R. (1987) Structure and transcription of the 5S rRNA gene from spinach chloroplasts. Curr. Genet. 12:263-269

Boczar B.A., Delaney T.P., and Cattolico R.A. (1989) Gene for the ribulose-1,5-bisphosphate carboxylase small subunit protein of the marine chromophyte *Olisthodiscus luteus* is similar to that of a chemo-autotrophic bacterium. Proc. Natl. Acad. Sci. USA 86:4996-4999.

Briat J.F., Lescure A.M., and Mache R. (1986) Transcription of the chloroplast DNA: a review. Biochimie 68:981-999.

Cattolico R.A., and Loiseaux-de Goër S. (1989) Analysis of chloroplast evolution and phylogeny: a molecular approach, pp. 85-100 *In*: J.C. Green, B.S.C. Leadbeater, and W.L. Diver (eds), The Chromophyte Algae: Problems and Perspectives. Clarendon Press, Oxford.

Christensen T. (1964) The gross classification of algae, pp. 59-64 *In*: D.F. Jackson (ed), Algae and Man. Plenum Press, New York.

Gibbs S.P. (1981) The chloroplasts of some algal groups may have evolved from endosymbiotic eukaryotic algae. Ann. N.Y. Acad. Sci. 361:193-207.

Gibbs S.P. (1990) The evolution of algal chloroplasts, pp. 145-157 *In*: D.G. Robinson and R.C. Star (eds), Experimental Phycology 1. Springer Verlag, New York.

Gunderson J.H., Elwood H., Ingold A., Kindle K., and Sogin M.L. (1987) Phylogenetic relationships between chlorophytes, chrysophytes and oomycetes. Proc. Natl. Acad. Sci. USA 84:5823-5827.

Kushel M., and Kowallik V. (1987) The plastome of a brown alga , *Dictyota dichotoma*. II. Mol. Gen. Genet. 207:361-368.

Lemieux B., Turmel M., and Lemieux C. (1988) Unidirectional gene conversions in the chloroplast of *Chlamydomonas* interspecific hybrids. Mol. Gen. Genet. 212:48-55.

Loiseaux-de Goër S., Markowicz Y., Dalmon Y., and Audren H. (1988) Physical maps of the two circular plastid DNA molecules of the brown alga *Pylaiella littoralis* (L.) Kjellm. Location of the rRNA genes and of several protein coding regions on both molecules. Curr. Genet. 14:155-162.

Manhart J.R., Kelley K., Dudock B.S., and Palmer J.D. (1989) Unusual characteristics of *Codium fragile* chloroplast DNA revealed by physical and gene mapping. Mol. Gen. Genet. 216:417-421.

Manhart J.R., and Palmer J.D. (1990) The gain of two chloroplast tRNA introns marks the green algal ancestors of land plants. Nature 345:268-270.

Markowicz Y., Mache R., and Loiseaux-de Goër S. (1988a) Sequence of the plastid rDNA spacer region of the brown alga Pylaiella littoralis (L.) Kjellm. Evolutionary significance. Plant Mol. Biol. 10:465-469.

Markowicz Y., Loiseaux-de Goër S., and Mache R. (1988b) Presence of a 16S rRNA pseudogene in the bi-molecular plastid genome of the primitive brown alga *Pylaiella littoralis*. Evolutionary implications. Current Genet. 14:599-608.

Perasso R., Baroin A., Qu L.H., Bachellerie J.P., and Adoutte A. (1989) Origin of the algae. Nature 339:142-144.

Raafat El-Gewely M., Helling R.B., Dibbits J.G. Th. (1984) Sequence and evolution of the regions between the *rrn* operons in the chloroplast genome of *Euglena gracilis* bacillus. Mol. Gen. Genet. 194:432-443.

Specht T., Wolters J., and Erdmann V.A. (1990) Compilation of 5S rRNA and 5S rRNA gene sequences. Nucl. Acids Res. 18(supplement):2215-2230.

Strittmatter G., and Kossel H. (1984) Cotranscription and processing of 23S, 4.5S and 5S rRNA in chloroplasts of *Zea mays*. Nucl. Acids Res. 12:7633-7647.

Woese C.R. (1987) Bacterial evolution. Microbiol. Rev. 51:221-271.

Yamada T., and Shimajii M. (1986) Peculiar feature of the organization of rRNA genes of the *Chlorella* chloroplast DNA. Nucl. Acids Res. 14:3827-3839.

THE SOLUTION STRUCTURE OF SPINACH CHLOROPLAST AND OF XENOPUS LAEVIS OOCYTE 5S rRNAs

P. Romby, C. Brunel, E. Westhof, F. Baudin, P.J. Romaniuk[1], R. Mache[2], C. Ehresmann and B. Ehresmann

Institut de Biologie Moléculaire et Cellulaire du CNRS, 15 rue R. Descartes, 67085 Strasbourg Cedex, France. [1]University of Victoria, Department of Biochemistry and of Microbiology, PO Box 1700, Victoria V8W 2Y2, Canada. [2]Laboratoire de Biologie Moléculaire Végétale, CNRS and Université de Grenoble I, 38041 Grenoble Cedex, France.

Summary

Detailed atomic models of a eubacterial 5S rRNA (from spinach chloroplast) and of a eukaryotic 5S rRNA (from Xenopus laevis oocyte) were built using computer graphic modeling. These models integrate the experimental data on the accessibility of bases and phosphates towards several structure-specific probes and the stereochemical constraints. The proposed 3-dimensional models adopt a Y-shaped structure with no tertiary interactions between the different domains of the RNA. The models have been corroborated by footprinting experiments with ribosomal proteins and by the analysis of various mutant 5S rRNAs. From these studies, several conclusions can be reached: (i) the two 5S rRNA molecules adopt a Y-shape structure with no long range tertiary interactions between the different helical domains. (ii) The internal loops, in particular region E, contain several non canonical base pairs of A-A, A-G and U-U types. The intrinsic structure of these loops are strictly sequence-dependent. (iii) The hinge nucleotides in loop A govern the relative orientation of the three helical domains.

Introduction

The ribosomal 5S RNA is a universal component of the large ribosomal subunit, except in mammalian mitochondria. Due to its small size, it has long been regarded as the system of choice for many kinds of investigations of rRNA structure and function. However, beyond its contribution to the structural integrity of the ribosomal 50S subunit, its role in protein synthesis still remains obscure. Extensive sequence comparison of 5S rRNAs from eukaryotic, eubacterial, archaebacterial and organelle ribosomes has led to the elaboration of a consensus secondary structure model with highly conserved sequence elements (for a compilation, see Wolters and Erdmann, 1988). Such a highly conserved structure in the course of evolution is consistent with a common role in ribosome function. Using a

NATO ASI Series, Vol. H 55
The Translational Apparatus of Photosynthetic Organelles
Edited by R. Mache et al.

wide variety of experimental approaches, a number of tentative tertiary models has been proposed, especially for Escherichia coli 5S rRNA (e.g. Böhm et al., 1981; Pieler and Erdmann, 1982; Hancock and Wagner, 1982; Göringer and Wagner, 1986). In several eukaryotic 5S rRNAs, tertiary interactions between loop C and loop D have been suggested (Toots et al., 1982; McDougall and Nazar, 1983). Otherwise, other models have been proposed in which the three domains are independent (Österberg et al., 1976; Fox and Wong, 1979; Christiansen and Garrett, 1976; Christiansen et al., 1987).

In a previous work, we have mapped at nucleotide resolution the conformation of spinach chloroplast 5S rRNA (Romby et al., 1988) and the Xenopus laevis oocyte and somatic 5S rRNAs (Romaniuk et al., 1988) using several structure-specific probes. From these structure-probing data, a three-dimensional model for the spinach chloroplast and X. laevis 5S rRNAs was constructed by computer graphic modeling integrating stereochemical constraints (Westhof et al., 1989). Since, neither stereochemical constraints nor the agreement between the experimental and calculated accessibilities can be taken as a proof, results independent from those introduced in the modeling should be used. The sites of interaction with other macromolecules obtained from footprinting data can give a useful indication as to the correctness of the model. However, the best criterion for judging the validity of the molecular model is the use of site-directed mutagenesis at selected residues considered critical for the tertiary folding of the RNA molecule. In this perspective, mutations have been introduced in the different loop regions of X. laevis 5S rRNA and their effect on the conformation of the RNA molecule has been checked by the probing approach (Baudin et al., 1990; Brunel et al., 1990; Leal de Stevenson et al., 1990; Romby et al., 1990a). In this paper, we have summarized the structural characteristics of spinach chloroplast 5S rRNA known as being of the eubacterial type and the eukaryotic 5S rRNA from X. laevis oocyte.

Experimental strategy

The probes

The conformation of the spinach chloroplast 5S rRNA and the somatic and oocyte X. laevis 5S rRNAs was tested by the use of several structure-specific probes. Prior to chemical modification or enzymatic hydrolysis, the 5S rRNA was carefully renatured in the presence of magnesium and KCl at pH 7,5. Single-strand specific nucleases (RNase T_1, nuclease S_1) and RNase V_1 specific for base-paired and structured regions were used. The four bases were tested at one of their Watson-Crick positions with dimethylsulfate (DMS) (at A(N1) and C(N3)) and with a carbodiimide derivative (CMCT) (at G(N1) and U(N3)). Position N7 of purines implicated in tertiary interactions, in base stacking or in cation coordination was tested with diethylpyrocarbonate (DEPC) (at A(N7)) and with DMS (at G(N7)). Ethylnitrosourea was used to identify phosphates involved in tertiary interactions through hydrogen bonds or in cation coordination. To estimate the degree of stability of helices, probing was achieved under native conditions (in the presence of magnesium and KCl, at 20°C) and under denaturing conditions (in the presence of EDTA, at 20°C). Two different methodological paths are then used to identify RNase cleavages and modified nucleotides. The enzymatic cleavages and the modified positions [G(N7) and C(N3) with DMS, A(N7) with DEPC, phosphates with ENU) were detected by using a 3' or 5' end-labeled 5S rRNA. The primer extension method was used to detect the bases modified by DMS [A(N1) and C(N3)] and by CMCT [U(N3) and G(N1)]. The experimental conditions were described in detail for chloroplast 5S rRNA by Romby et al. (1988) and for X. laevis 5S rRNA by Romaniuk et al. (1988). Advantages and limitations of the probing approach have been discussed by Ehresmann et al. (1987).

Graphic modeling

The modeling was done with the help of several programs as described by Westhof et al. (1989). With the program FRAGMENT, one can choose any structural motif existing in the structure data bank and insert along the sugar-phosphate backbone the desired base sequence with the appropriate numbering (Westhof, 1988). Standard RNA helices are constructed with the program NAHELIX. The different modeled RNA fragments are then assembled manually with the interactive program FRODO (Jones, 1978) adapted for Evans and Sutherland PS300. To ensure proper geometry for base pairing and correct stereochemistry, the output of FRODO is subjected to Hendrickson-Konnert restrained least squares with the program NUCLIN and NUCLSQ (Konnert and Hendrickson, 1980; Westhof et al., 1985). At the end as well as during the construction of the sub-structures, a quantitative measure of the accessibility of each atom within the tertiary model is performed with a program based on the algorithm and definition of Richmond (1984). The resulting values are then compared to the experimental results and the model is corrected until to get a good correlation between the calculated accessibilities and the chemical reactivities.

The conformation of spinach chloroplast and Xenopus laevis oocyte 5S rRNAs

For the spinach chloroplast 5S rRNA and X. laevis oocyte and somatic 5S rRNAs, the experimental data support the existence of five helices proposed in the consensus secondary structure model (see Fig. 1). Helix V only contains three base pairs in chloroplast 5S rRNA and seven in Xenopus laevis 5S rRNA. The results also suggest the presence of bulged residues in helices II and III. In the internal and external loops, several nucleotides are clearly not reactive at Watson-Crick positions and at N7, in particular in loops C and E (Romby et al., 1988; Romaniuk et al., 1988). This suggests their involvement either in local intrinsic conformation or in long-range tertiary interactions. However, the stereochemical constraints imposed during the construction of the molecular model (Westhof et al.,

1989) favor the existence of independent domains in the 5S rRNA. This has been further supported by the study of the conformation of several mutant 5S rRNAs from X. laevis (Baudin et al., 1990; Brunel et al., 1990; Leal de Stevenson et al., 1990). Thus, on the basis on the chemical reactivities, several local interactions involving non canonical base pairs (G-U, A-A, A-G, or U-U) have been postulated for the three 5S rRNAs (see Fig. 1).

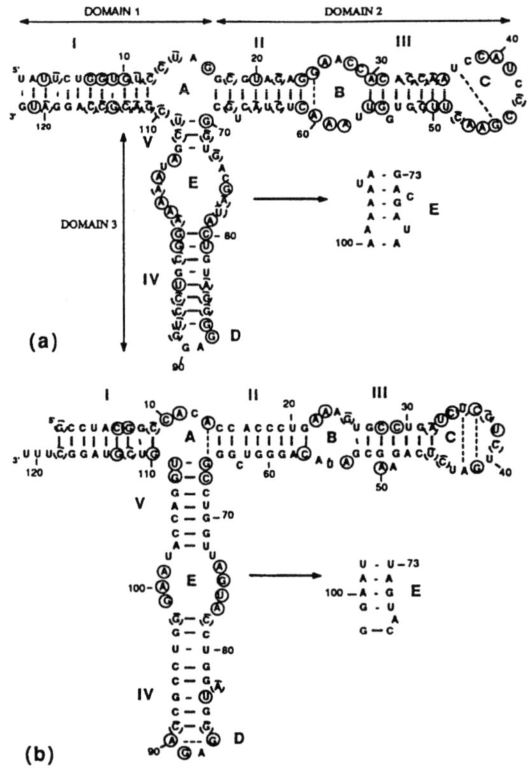

Figure 1 : Secondary structure of spinach chloroplast 5S rRNA (a) and of *Xenopus laevis* 5S rRNA (b). Helices and loops are numbered according to Romaniuk (1985). Nucleotides phylogenetically conserved in chloroplast 5S rRNAs (a) and in eukaryotic 5S rRNAs (b) are indicated: (O) strictly conserved nucleotides and (◌) semi-conserved nucleotides (adapted from Specht *et al.* ,1990).

A comparison of the data obtained for the two types of X. laevis 5S rRNAs indicates that the conformation of the oocyte 5S rRNA and that of the somatic 5S rRNA are very similar. Only, helices III and IV are less stable in the oocyte 5S rRNA and loop B adopts a more organized structure in the somatic 5S rRNA (Romaniuk et al., 1988; Westhof et al., 1989). Therefore, in what follows the discussion will be conducted on the spinach chloroplast 5S rRNA and the oocyte X. laevis 5S rRNA.

The tertiary models of both 5S rRNAs are shown in Figure 2(a,b). The two models adopt a distorted Y-shape with a short stalk made of helix I and with the two arms of the Y made of domains 2 and 3. The Y-shape is in the plane and the thickness is essentially that of a RNA double helix. The two models follow the secondary structure and among the published models ressemble those proposed

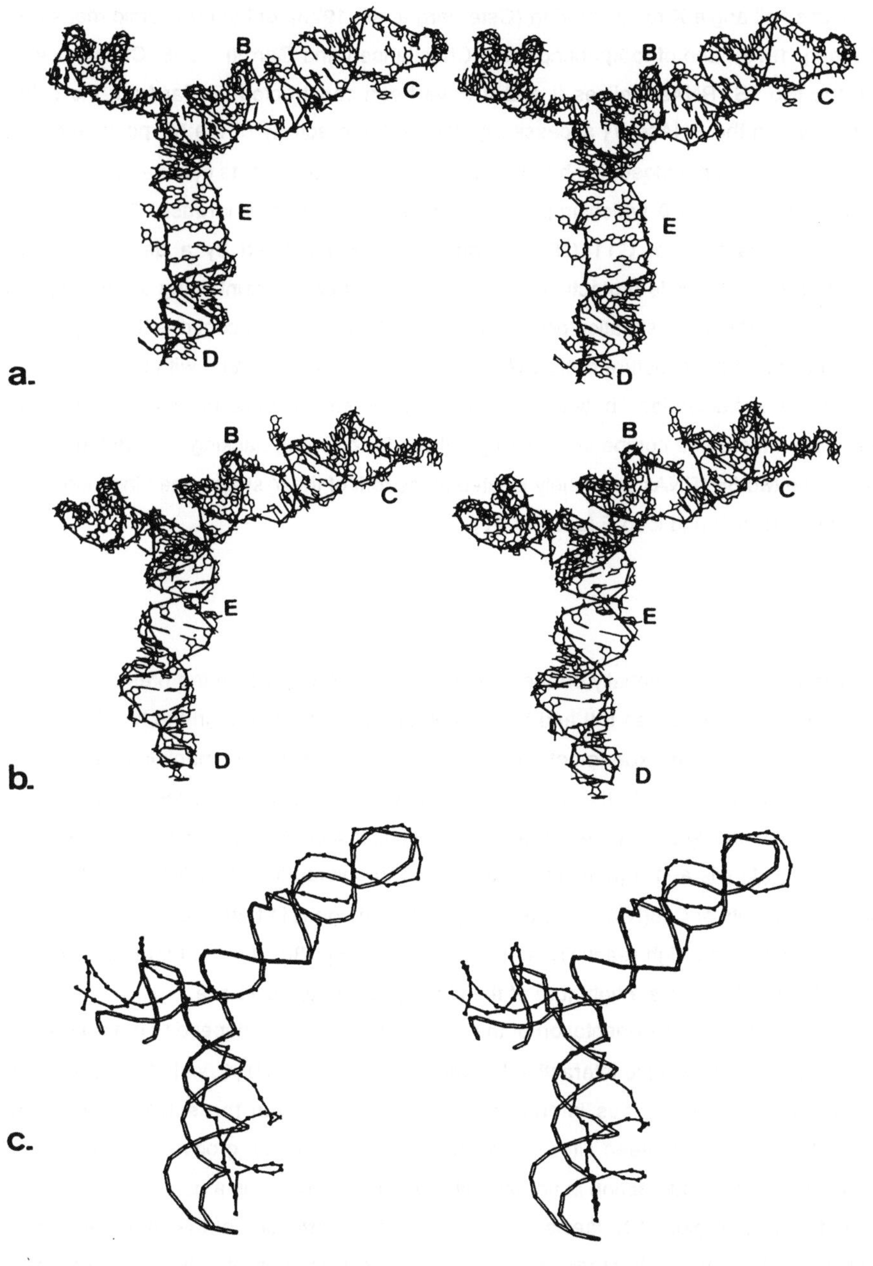

Figure 2 : Three-dimensional model of spinach chloroplast 5S rRNA (a) and of X. laevis 5S rRNA (b). The views are from Westhof et al. (1989). In (c) are superimposed the phosphate backbone of X. laevis (black line) 5S rRNA with the atomic view of the chloroplast 5S rRNA (white bond). The stereoviews are drawn with the program PLUTO (S. Motherwell and P. Evans, MRC Cambridge). The loops are numbered.

on the basis of small angle X-ray scattering (Österberg et al., 1976), of hydrodynamic measurements (Fox and Wong, 1979), and of footprinting data (Christiansen and Garrett, 1986; Christiansen et al., 1987). For the two 5S rRNAs, helices II and V have been constructed not far from the colinearity essentially based on the RNAse V_1 accessibility and on the reactivity of the N7 position of guanine located at the hinge region. Indeed, in chloroplast 5S rRNA, RNase V1 cuts occur almost continuously between nucleotides 5 to 30 which implies the absence of steric hindrance along this stretch. Furthermore, the reactivity of G111(N7) in chloroplast is explained by a destacking between nucleotides 110 and 112, while protection at G70(N7) is attained by maintaining a stacking between residues 69 and 70. The same situation occurs in X. laevis 5S rRNA. This confirms the empirical rule of RNA folding, based on the structure of tRNAs, according to which the strand with the least number of unpaired connecting nucleotides between helical sections should become part of the continuous double helix when there is a competition among helices for colinear stacking. The differences which exist between the two 5S rRNAs are mainly located at the fork of the Y-shape (the hinge loop A) and in region comprising helix V and loop E (see Fig. 2c).

The hinge region

The tertiary fork depends essentially on the conformation of loop A and on the presence of the hinge residues at the junction of helices I, II and V. The conformation is different in the two 5S rRNAs (see Fig. 3) as evidenced by the experimental data (Romby et al., 1988; Romaniuk et al., 1988). The differences are essentially due to the presence of an additional residue between helices I and V in chloroplast 5S rRNA and by the nature of the last nucleotide in loop A (it is a strictly conserved adenine at position 13 in X. laevis and a guanine at position 16 in chloroplast 5S rRNAs). In chloroplast 5S rRNA, the non-reactivity of C13(N3) is explained by a stacking on G111 while C110 protrudes towards the solvant explaining the high reactivity at its N3 position (Fig. 3b). Furthermore, the protection of phosphates 16 and 17 requires a tight turn of the sugar-phosphate backbone.

In X. laevis, the relative orientation of the three helical domains depends on a precise array of hydrogen bonds based on a triple interaction between A13, G66 and U109, with A13 occupying a key position that stabilizes the various interactions (see Fig. 3c). From the solution data, a tertiary interaction between the Hoogsteen sites of A13(N7,N6) and the minor groove sites of G66(O2',N3) was constructed. Such an interaction is not favorable when position 13 is a guanine. A series of point mutations in the hinge region of X. laevis 5S rRNA on the conformation of the RNA has been tested (Baudin et al., 1990). From this study, it appears that the formation of a Watson-Crick base pair between residues 13 and 66 not only prevents base pairing between residues 66 and 109, but also destabilizes the next G-C pairs in helix V. However, an alternative non-canonical base pairing may replace the original A13-G66 interaction without affecting the formation of the G66-U109 pairing. Interestingly, none of the studied mutations appears to change fundamentally the stacking rules between helices II and V. Only, the orientation between domains II and III is slightly perturbed (Baudin et al., 1990).

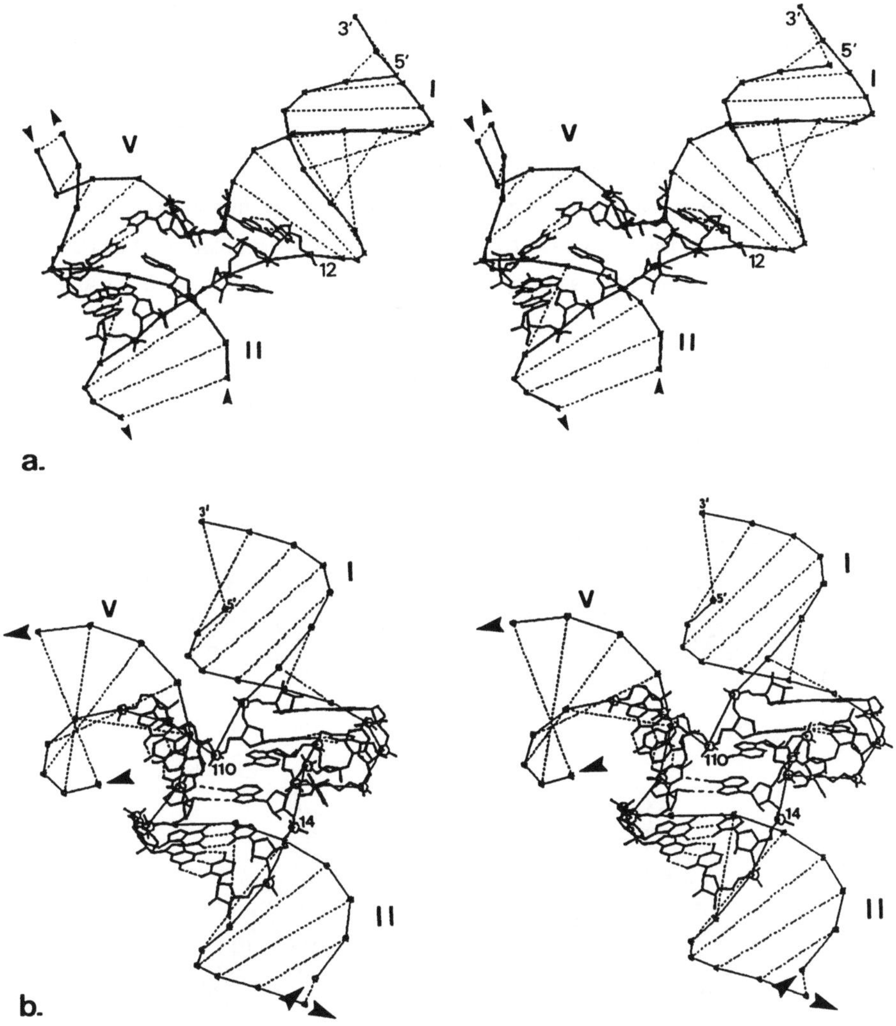

Figure 3 : Stereo views of the hinge region in (a) spinach chloroplast 5S rRNA (C8-G14, C69-G70, U109-G111) and (b) in X. Laevis 5S rRNA (G8-C15, G64-C67, G108-U111). The views are given in the same orientation. The helices are numbered.

Domain 2

Domain 2 contains helices II and III separated by the internal loop B and closed by the hairpin loop C. In the two 5S rRNAs, helix II is a regular 8 base pair-helix with a bulged pyrimidine at an analogous position. Helix III contains 6 base pairs with two bulged adenines in X. laevis 5S rRNA and contains 7 base pairs with one bulged guanine in chloroplast 5S rRNA. In loop B, most of the nucleotides are accessible towards the chemical probes in the two 5S rRNAs. In the chloroplast 5S rRNA, a G25-A60 base pair is proposed to explain the non-reactivity of G25 at position N1 and A60 at N7. Also, the non reactivity of the two adenines 26 and 27 is explained by a stacking effect. In X. laevis 5S rRNA, loop B adopts a more disorganized structure in which most of the bases are exposed towards the solvant with the presence of a U26-G53 base pair. The rather unconstrained structure of loop B probably allows a certain degree of movement of the helix III/loop C region.

Loop C contains twelve residues with several invariant positions. Interestingly, in the two loops the possibility exists to form two base pairs (U37-A46 and C38-G45 in chloroplast or U33-A42 and C34-G41 in X. laevis), leading to a six base loop motif. However, the probing data do not support such

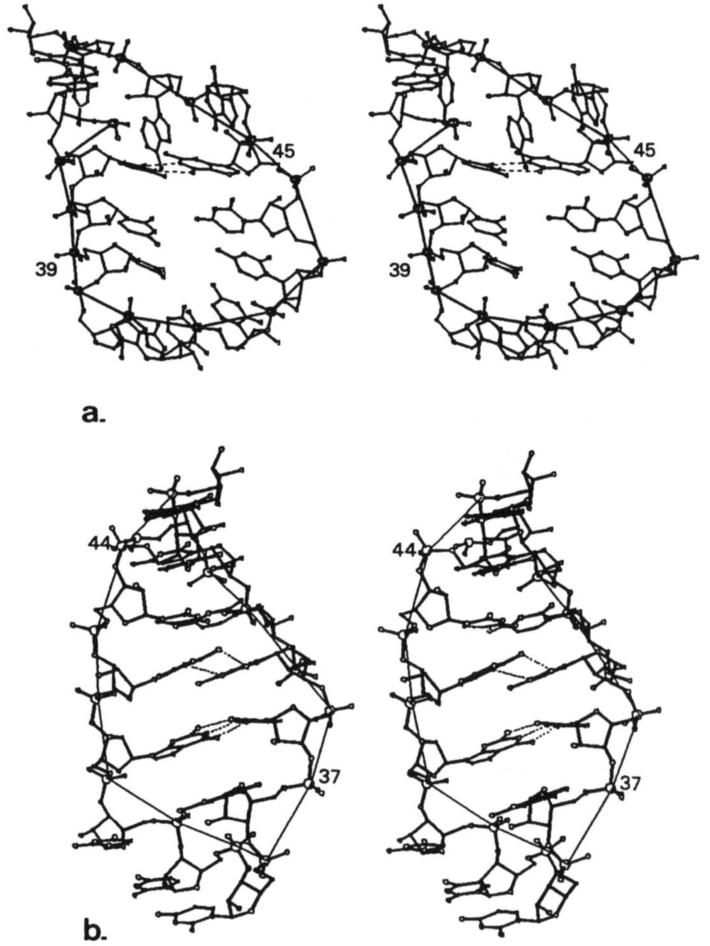

Figure 4 : Stereo views of loop C in (a) spinach chloroplast 5S rRNA (A36-U49) and in (b) X. laevis 5S rRNA (A32-U45). Views are from Westhof et al. (1989).

interactions. Indeed, the residues in loop C of X. laevis 5S rRNA are less accessible towards the chemical probes (with U33 to C36, G41, U43 and C44 unreactive at Watson-Crick position, and A42 at N7) than in chloroplast 5S rRNA (where only U37 at N3 and G45 at N1 are non-reactive under native conditions). The resulting conformation of the two loops are shown in Figure 4. In chloroplast 5S rRNA, a G45-U37 base pair is proposed with A46 to C48 bulging out in an orientation roughly parallel to the helical axis of helix III (Fig. 4). The conformation of the seven residues left was constructed following the same principles as for the anticodon loop of the tRNA. The turn is made at C39 and residues A40 to C42 are orientated towards the solvant, the loop being closed in helical continuity by residues C43 to G45. In X. laevis 5S rRNA, the pyrimidine residues (U33, C34, U43 and C44) are stacked inside the loop but without hydrogen bonding, explaining their non reactivity at N3. Two base pairs follow, U35-A42 in a trans-Hoogsteen arrangement and C36-G41 in a Watson-Crick arrangement. This leads to the formation of a 4-membered loop motif with the turn at G37 and the three bases 38 to 40 pointing towards the major groove of helix III (Fig. 4).

The effect of mutations in loop C on the conformation of X. laevis 5S rRNA has been studied (Brunel et al., 1990; Romby et al., 1990b). The main result is that the overall conformation of the 5S rRNA is not affected by the mutations while conformational changes primarily occured in the mutated loop C. This study clearly demonstrates the absence of tertiary interaction between loop C and other regions of the RNA. It also shows that (i) the nucleotides U35, C36, G41 and A42 play a crucial role in maintaining an intrinsic conformation in loop C of X. laevis 5S rRNA; (ii) the sequence has a direct effect on the flexibility and on the intrinsic conformation of loop regions; (iii) also, it appears that nucleotides in large loops tend to fold into organized conformations, forming structural motives that probably increase the stability (Brunel et al., 1990).

Domain 3

Domain 3 contains helices IV and V separated by the region E (given as an internal loop in the consensus model, see Fig. 1) and terminated by loop D. In the two 5S rRNAs, helix IV contains seven base pairs and is closed by a four base loop. In X. laevis 5S rRNA, the probing data support the existence of a U80-U96 base pair and a bulged residue involving a dynamic equilibrium between A83 and U84 in helix IV. In chloroplast 5S rRNA, the four purines in loop D are reactive at both N1 and N7 positions (Romby et al., 1988). Therefore, loop D was constructed with G88 and A89 on one side of the loop and G90 and G91 on the other side, with the phosphate 91 interacting with O2' of the ribosyl moiety of residue 89 (Westhof et al., 1989). The sequence motif GAGA found in X. laevis 5S rRNA is highly conserved in eukaryotic 5S rRNA and prokaryotic 16S rRNA. Interestingly, the same chemical reactivity pattern has been observed in a number of four-membered hairpin loops of the same sequence (GAGA) in E. coli 16S rRNA (Baudin et al., 1987) which strongly suggests that the loop adopts a particular conformation. The adenine residues are reactive at N1 and the two guanines unreactive at both N1 and N7 positions (Romaniuk et al., 1988). Although other possibilities cannot be excluded, the formation of a G87(N2,N3)-A90(N7,N6)syn bond and a turn at A88 may account for most of the observed reactivity pattern.

The most striking difference between the two 5S rRNAs is related to the conformation of the helix V/loop E region. Indeed, in chloroplast 5S rRNA helix V contains three base pairs with a large loop E while in X. laevis 5S rRNA helix V contains seven base pairs and a short loop E. However, in the two 5S rRNAs, loop E appears to be highly organized. The loop is found to be resistant to both single-strand and double-strand specific RNases, all the purines are unreactive at their N7 position and several phosphates have been found protected (Romby et al., 1988; Romaniuk et al., 1988). In X. laevis 5S rRNA, we have probed the conformation of a fragment (obtained from cloned gene by in vitro transcription with T_7 RNA polymerase) which contains helices IV and V, loop E and loop D. This fragment adopts exactly the same conformation as the corresponding region in the entire 5S rRNA molecule (Leal de Stevenson et al., 1990). This study clearly demonstrates that the three domains are independent and that loop E adopts an intrinsic structure. Therefore, based on the probing data, the structure which was proposed for loop E in chloroplast 5S rRNA contains several unusual A(N6,N7)-A (N7,N6) in a syn orientation and A(N6,N7)syn-G(O6,N1)anti base pairs and three bulged residues C75, U78, and U105 (Romby et al., 1988; Westhof et al., 1990). Due to the presence of A79-A100 and A77-A102 base pairs, it was not possible to construct a Watson-Crick base pair between U78 and A101. Therefore, U78 is bulging out and A101 is stacked inside the helix, thus explaining the high reactivity of U78 at N3 and A101 at N1, and the non reactivity of A101 at N7. The non reactivity of phosphates 76 and 107 might be explained by the assumption of cation binding (such as magnesium) that could help to stabilize the unusual phosphate backbone conformation. In this model, it appears that the two RNA strands do not display the feature of a standard A-helix but rather adopt an unwound and extended configuration (Fig. 5a). Structure conformations have been postulated for loop E in E. coli (Gewirth et al., 1987; Zhang and Moore, 1989), different from that of chloroplast 5S rRNA despite their strong sequence homologies (Zhang and Moore, 1989). This observation most probably indicates the importance of nucleotide sequence on the intrinsic conformation of loop E.

In X. laevis, several non canonical interactions similar to those found in chloroplast 5S rRNA, albeit over fewer nucleotides, have also been proposed (Romaniuk et al., 1988; Westhof et al., 1989). This structure contains a U73(O4,N3)-U102(N3,O2) followed by a A74-A101 pair connected via their N6 and N7 atoms, and a G75(O6,N1)anti-A100(N6,N7)syn (see Fig. 5b and Westhof et al., 1989), according to the probing data. Adenine 77 non reactive at N7, was stacked inside the helix forming a slight wedge. Again, the protection of phosphates 75 and 100 can be explained by the presence of a magnesium binding site. In this model, loop E adopts a helix geometry that is only slightly distorted from the A-type, and more regular than that found in chloroplast 5S rRNA (Fig. 5b). This difference accounts for the globally different aspect of domain 3 in the two 5S rRNAs (Fig. 2c). Several mutations in loop E have been tested on the conformation of X. laevis 5S rRNA (Leal de Stevenson et al., 1990). The results indicate that base substitutions on either side of loop E result in a collapse of the local conformation to an unstructured loop. Also, the stability of the base pairs closing the loop in helices IV

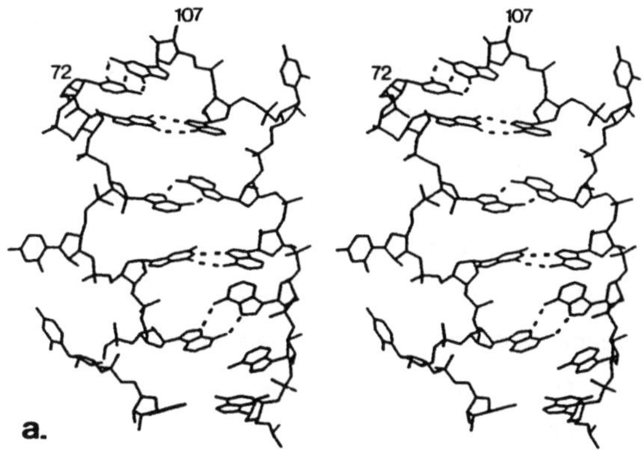

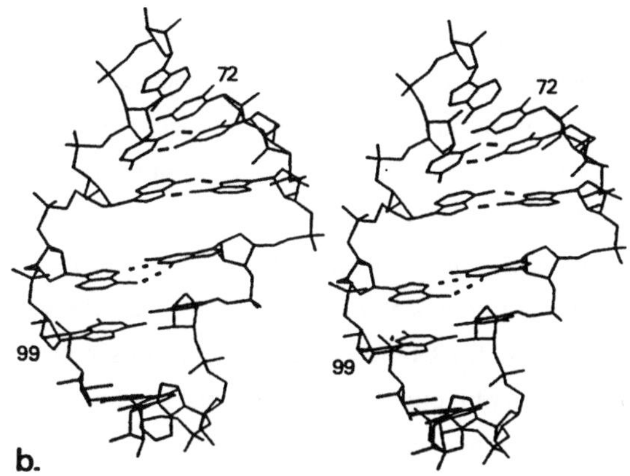

Figure 5 : Stereo models of loop E in (a) spinach chloroplast 5S rRNA (U72-C80/G99-G107) and in (b) X. laevis 5S rRNA (U72-C78/G98-A103). The views are from Romby et al. (1988) and Westhof et al. (1989), respectively. The hydrogen bonds are shown by broken lines.

and V is affected when the mutations disrupt the unusual conformation of the loop. From this study, it also appears that the intrinsic loop conformation is strictly sequence-dependent and that magnesium is important to stabilize the particular folding of loop E. These mutations do not affect the conformation of the other domains of the 5S rRNA, showing again the absence of long-range tertiary interactions between domains 2 and 3.

Conclusion

The goal of the present paper was to summarize the criteria used to construct and compare the tertiary structure model of the two different 5S rRNA structures, one from eubacterial type (chloroplast 5S rRNA) and one from eukaryotic type (X. laevis oocyte 5S rRNA). The value of the models lies in a combination of a variety of different approaches (chemical probing, phylogenetic comparison, protein binding, mutant 5S rRNAs) into a structure that makes sense in terms of design and stability. It is interesting to note that the model can be generalized to the prokaryotic and eukaryotic 5S rRNAs. The two models adopt a Y-shaped structure with helices II and V not far from coaxiality. The main differences which exist between the two 5S rRNAs are located in the hinge region (loop A) and in helix V/loop E region. An extensive study of the conformation of several mutant 5S rRNAs from X. laevis has clearly shown that the three domains of the RNA are independent and that the loop regions adopt intrinsic conformation mainly based on non canonical interactions. This study stresses the importance of the sequence dependence of intrinsic conformations and their role in the flexibility of these regions. In the two models, several non canonical interactions are present in the loop regions. It appears that these particular base pairs are widely distributed in the RNA molecules. They influence the geometry of the RNA molecules and should provide specific signals for protein recognition. In spinach chloroplast ribosomes, two proteins that bind specifically and independently to the 5S rRNA have been characterized (Toukifimpa et al., 1989). The binding sites of the two proteins do not overlap on the RNA molecule and are well explained by the proposed tertiary model of the 5S rRNA (Toukifimpa et al., 1989; Mache et al., in this issue). The binding sites of E. coli ribosomal proteins L18 and L25 also agree with the Y-shape structure. Indeed, L18 was essentially found centered on helix II/loop B/helix III (Christiansen et al., 1985; Edjeberg et al., 1989), while L25 recognized a fragment containing helices V/loop E/helix IV (Kime et Moore, 1983). Also, the coaxiality between helices II and V has been suggested for X. laevis 5S rRNA to explain the binding site of transcription factor TFIIIA (Christiansen et al., 1987). Furthermore, our study provides insights into the general problem of RNA folding in bifurcation regions. In chloroplast 5S rRNA, the colinearity between helices II and V is mainly maintained by stacking interaction while in X. laevis 5S rRNA the orientation of the helices depends on a precise array of hydrogen bonds based on the triple interaction between A13, G66 and U109 (which are highly conserved in eukaryotic 5S rRNA).

One important question that should be addressed concerns the conformation of the 5S rRNA within the ribosome. In the body of the large subunit, does 5S rRNA adopt the same conformation as that found for the naked 5S rRNA in solution. Nevertheless, it has to be noticed that the localisation of the binding sites of the ribosomal proteins agree well with the Y-shape model (e.g. Christiansen et al., 1985, 1987; Edjeberg et al., 1989; Toukifimpa et al., 1989). In fact, only limited information is available on the structure of 5S rRNA within the ribosome. It appears that 5S rRNA is tightly embedded in the subunit structure, leaving a few accessible sites (Herr and Noller, 1974; Farber and Cantor, 1981).

References

Böhm, S., Fabian, H., Venyaninov, S.Y., Matveer, S.V., Lucius, H., Welfe, H. and Filimonov, V.V. (1981) Structural analysis of the A and b conformaers of Escherichia coli 5S ribosomal RNA by infrared spectroscopy. FEBS Lett. 132: 357-361.

Baudin, F., Ehresmann, C., Romby, P., Mougel, M., Colin, J., Lempereur, L., Bachellerie, J.P., Ebel, J.P. and Ehresmann, B. (1987) Higher-order structure of domain III in Escherichia coli 16S ribosomal RNA, 30S subunit and 70S ribosome. Biochimie 69, 1081-1096.

Baudin, F., Romaniuk, P.J., Romby, P., Brunel, C., Westhof, E., Ehresmann, B. and Ehresmann, C. (1990) Involvement of "hinge" nucleotides of X. laevis 5S rRNA in the RNA structural organization and in the binding of transcription factor TFIIIA. J. Mol. Biol., in press.

Brunel, C., Romby, P., Westhof, E., Romaniuk, P.J., Ehresmann, B. and Ehresmann, C. (1990) Effect of mutations in domain 2 on the structural organization of oocyte 5S rRNA from X. laevis . J. Mol. Biol. 215: 103-111.

Christiansen, J. and Garrett, R.A. (1986) In "Structure, Function and Genetics of Ribosomes" (Hardesty, B. and Kramer, G., eds.), Springer Verlag, pp 253-269.

Christiansen, J., Brown, R.S., Sproat, B.S. and Garrett, R.A. (1987) Xenopus transcription factor IIIA binds primarily at junctions between double helical stems and internal loops in oocyte 5S RNA. EMBO J. 6: 453-460.

Egebjerg, J., Christiansen, J., Brown, R.S., Larsen, N. and Garrett, R.A. (1989) Protein L18 binds primarily at the junctions of helix II and internal loops A and B in Escherichia coli 5S RNA. J. Mol. Biol. 206: 651-668.

Ehresmann, C., Baudin, F., Mougel, M., Romby, P., Ebel, J.P. and Ehresmann, B. (1987) Probing the structure of RNA in solution. Nucleic Acids Res. 15: 9109-9128.

Farber, N.M. and Cantor, C.R. (1981) A slow tritium exchange study of the solution structure of Escherichia coli 5S ribosomal RNA. J. Mol. Biol. 146: 241-257.

Fox , J.W. and Wong, K.P. (1979) The hydrodynamic shape, conformation, and molecular model of Escherichia coli ribosomal 5S rRNA. J. Biol. Chem. 254: 10139-10144.

Gewirth, D.T. and Moore, P.B. (1987) Secondary structure of 5S rRNA: NMR experiments on RNA molecules partially labeled with nitrogen-15. Biochemistry 26: 5657-5665.

Göringer, H.U. and Wagner, R. (1986) Does 5S RNA from Escherichia coli have a pseudoknotted structure? Nucleic Acids Res. 14: 7473-7485.

Hancock, J. and Wagner, R. (1982) A structural model of 5S RNA from Escherichia coli based on intramolecular crosslinking evidence. Nucleic Acids Res.10: 1257-1269.

Herr, W. and Noller, H.F. (1979) Protection of specific sites in 23S and 5S RNA from chemical modification by association of 30S and 50S ribosomes. J. Mol. Biol. 130, 421-432.

Jones, T.A. (1978) A graphic model building and refinement system for macromolecules. J. Appli. Crystallogr. 11: 268-278.

Kime, M.J. and Moore, P.B. (1983) Nuclear overhauser experiments at 500 MHz on the downfield proton spectra of 5S ribonucleic acid and its complex with ribosomal protein L25. Biochemistry 22, 2622-2629.

Konnert, J.H. and Hendrickson, W.A. (1980) A restrained-parameter thermal-factor refinement procedure. Acta Crystallogr. Sect. A 36: 344-349.

Leal de Stevenson, I., Baudin, F., Brunel, C., Romby, P., Ehresmann, C., Ehresmann, B. and Romaniuk, P.J. (1990) Structural studies on site-directed mutants of domain 3 of Xenopus laevis oocyte 5S ribosomal RNA. Submitted.

McDougall, J. and Nazar, R.N. (1983) Tertiary structure of the eukaryotic ribosomal RNA: accessibility of phosphodiester bonds to ethylnitrosourea modification. FEBS Lett. 258: 5256-5259.

Österberg, R., Sjoberg, B. and Garrett, R.A. (1976) Molecular model for 5S rRNA: a small-angle X-ray scattering study of native, denatured and aggregated 5S RNA from Escherichia coli ribosomal 5S RNA. Proc. Natl. Acad. Sci. U.S.A. 77: 4679-4682.

Pieler, T. and Erdmann, R.A. (1982) 3-Dimensional structural model of eubacterial 5S rRNA that has functional implications. Proc. Natl. Acad. Sci. U.S.A. 79: 4599-4603.

Richmond, T.J. (1984) Solvent accessible surface area and excluded volume in proteins. J. Mol. Biol. 178: 63-89.

Romaniuk, P.J. (1985) Characterization of the RNA binding properties of transcription factor IIIA of Xenopus laevis oocytes. Nucleic Acids Res. 13: 5369-5387.

Romaniuk, P.J., Leal de Stevenson, I., Ehresmann, C., Romby, P. and Ehresmann, B. (1988) A comparison of the solution structures and conformational properties of the somatic and oocyte 5S rRNA of Xenopus laevis. Nucleic Acids Res. 16: 2295-2312.

Romby, P., Westhof, E., Toukifimpa, R., Mache, R., Ebel, J.P., Ehresmann, C. and Ehresmann, B. (1988) Higher-order structure of chloroplastic 5S ribosomal RNA from spinach. Biochemistry 27: 4721-4730.

Romby, P., Baudin, F., Brunel, C., Leal de Stevenson, I., Westhof, E., Romaniuk, P.J., Ehresmann, C. and Ehresmann, B. (1990a) Ribosomal 5S RNA from Xenopus laevis oocytes: conformation and interaction with transcription factor IIIA. Biochimie 72: 437-452.

Romby, P., Brunel, C., Westhof, E., Ehresmann, C. and Ehresmann, B. (1990b) In "Water and Ions in Biomolecular Systems" (Vasilescu, D., Jaz, J., Packer, L. and Pullman, B., eds.), Birkhäuser press, Verlag Basel, pp19-27.

Specht, T., Wolters, J. and Erdmann, V. (1990) Compilation of 5S rRNA and 5S rRNA gene sequences. Nucleic Acids Res. 18: 2215-2230.

Toots, I., Misselwitz, R., Böhm, S., Welfe, H., Villems, R. and Saarma, M. (1982) Two distincts conformations of rat liver ribosomal rRNA. Nucleic Acids Res. 10: 3381-3389.

Toukifimpa, R., Romby, P., Rozier, C., Ehresmann, C., Ehresmann, B. and Mache, R. (1989) Characterization and footprinting analysis of two 5S rRNA binding proteins from spinach chloroplast ribosomes. Biochemistry 28: 5840-5846.

Westhof, E., Dumas, P. and Moras, D. (1985) Crystallographic refinement of yeast aspartic acid transfer RNA. J. Mol. Biol. 184: 119-145.

Westhof, E. (1988) Computer-aided structural biochemistry of nucleic acid. In "Computer Aided Molecular Design" IBC Technical Services, Ltd. London.

Westhof, E., Romby, P., Romaniuk, P.J., Ebel, J.P., Ehresmann, C. and Ehresmann, B. (1989) Computer modeling from solution data of spinach chloroplast and of Xenopus laevis somatic and oocyte 5S rRNAs. J. Mol. Biol. 207: 417-431.

Wolters, J. and Erdmann, V.A. (1988) Compilation of 5S rRNA and 5S rRNA gene sequences. Nucleic Acids Res. 16 (Suppl.): r1-r70.

Zhang, P. and Moore, P.B. (1989) An NMR study of the helix V-loop E region of 5S RNA from Escherichia coli. Biochemistry 28: 4607-4615.

Chloroplast tRNAs and tRNA genes : structure and function

L. Maréchal-Drouard, P. Guillemaut, H. Pfitzingzer and J.H. Weil
Institut de Biologie Moléculaire des Plantes du CNRS, Université Louis Pasteur
12 rue du Général Zimmer
67084 Strasbourg-cedex
France.

I. Organization and number of chloroplast tRNA genes

Chloroplast (cp) genomes of three land plants, namely tobacco, rice and *Marchantia polymorpha*, have now been completely sequenced. In tobacco (Fig. 1), the cp genome codes for 30 tRNA species (Shinozaki et al. 1986 ; Sugiura and Wakasugi 1989), 23 being encoded by single genes (located in the large single copy (LSC) or small single copy (SSC) regions), 7 being encoded by genes present twice (in the IR sequences). In rice the same set of tRNA genes was found (Hiratsuka et al. 1989). However, a total of 31 tRNA species are encoded by the liverwort *Marchantia polymorpha* cp genome ; the additional tRNA gene (with respect to tobacco and rice) is a $tRNA^{Arg}$(CCG) gene (Ohyama et al., 1986).
Pseudogenes have been found in a number of cp genomes. For instance, a pseudo-$tRNA^{Ile}$ and a pseudo-$tRNA^{Trp}$ genes (El-Gewely et al. 1984) have been found in *Euglena gracilis* strain *bacillaris*. A pseudo-$tRNA^{Pro}$ gene with a GGG anticodon has been sequenced in liverwort (Ohyama et al. 1986). A $tRNA^{fMet/Gly}$ chimeric structure has been described in wheat (Howe 1985) and in rice (Hiratsuka et al. 1989). A pseudo-$tRNA^{Thr}$ gene has been reported in barley (Oliver and Poulsen 1984), wheat (Quigley and Weil 1985) and rice (Hiratsuka et al. 1989).

NATO ASI Series, Vol. H 55
The Translational Apparatus of Photosynthetic Organelles
Edited by R. Mache et al.

II. Intervening sequences in chloroplast tRNA genes

No intron has been found in the *Euglena gracilis* cp tRNA genes, whereas 6 tRNA genes are interrupted in the case of higher plant (Fig. 1) and liverwort cp genomes (Ohyama et al. 1986 ; Shinozaki et al. 1986 ; Hiratsuka et al. 1989).

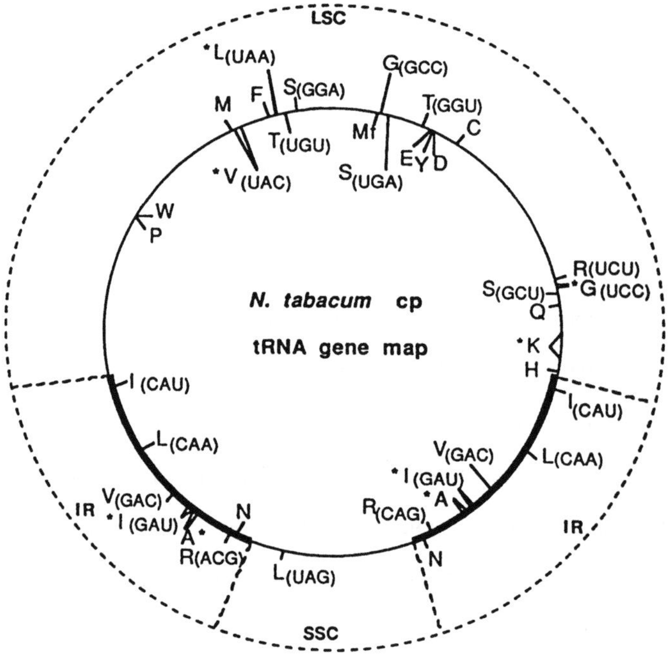

Figure 1 : tRNA gene organization of the chloroplast genome of *Nicotiana tabacum*. The inverted repeats (IR) are shown in bold lines. LSC and SSC indicate the large and small single copy region, respectively. The tRNA genes are shown by the one-letter code for amino acids with their anticodons in brackets (when several isoaccepting species are present). Asterisks indicate split genes. Genes shown on the outside of the circle are transcribed counterclockwise ; those inside are transcribed clockwise.

These introns which are very long as compared to the size of the mature tRNA (from 325 to 2526 bp) must be spliced efficiently since the corresponding mature tRNAs are found in relatively high amounts in chloroplasts. The introns found in the $tRNA^{Ala}$, $tRNA^{Gly}$(UCC), $tRNA^{Ile}$(GAU), $tRNA^{Lys}$ and $tRNA^{Val}$(UAC) genes belong to the class-II of structured introns, whereas the intron found in the $tRNA^{Leu}$(UAA) gene belong to the class-I of structured introns (Michel and Dujon 1983). Interestingly, the latter intron is the only class-I intron found in land plant cp genomes up to now and a similar intron was also found in the corresponding gene in the cyanelles of *Cyanophora paradoxa* (Evrard et al. 1988), but not in the chloroplasts of *Euglena gracilis*. The class-II introns found in the $tRNA^{Ala}$ and $tRNA^{Ile}$(GAU) genes are present in all land plants studied so far, but not in other organisms (Markowicz et al. 1988). All the cp tRNA gene introns are located in the anticodon loop, except that of the $tRNA^{Gly}$(UCC) gene where the intron is located in the D stem of the tRNA.

III. Chloroplast tRNA gene promoters

Comparison of the 5'-flanking sequences of land plant cp tRNA genes, as well as of protein genes, have revealed the presence of prokaryotic-like promoters upstream of these genes. As illustrated in figure 2 in the case of *trnE*, two conserved motifs similar to the bacterial "-35" and "-10" boxes can be identified upstream of this gene in land plant chloroplasts and in the cyanelles of *Cyanophora paradoxa*. That these motifs represent the actual promoter elements of this gene is also suggested by S1 mapping experiments which demonstrated that a precursor transcript starts 8 bp downtream of the "-10" element (Ohme et al. 1985). However, proof of the involvement of such motifs in the transcription of cp tRNA genes have up to now only been obtained in the case of spinach $tRNA^{Met}_2$ by expression of the corresponding gene *in vitro* in a homologous soluble transcription system : In this case mutations in both the "-35" and "-10" boxes cause a decrease of transcription (Gruissem and Zurawski 1985a).

```
                "-35"                      "-10"
N.t. CATTCGATTATATTGACAATTTCAAAAAACTGATCATACTATGATCATAGTATGATGGCG..13
V.f. CATTCCATTATATTGACAATTTCAAAAAACTGTTCATACTATGAACATAGTAGAATGGAA..13
P.s. CGAATCATTATATTGACAATTTCAAAAAACTGTTCATACTATGAACATAGTAGAATGGAG..13
S.o. TATTCGATTATATTGACAATTTCAAAAAACTGTTCATACTATGAACATAGTATGATGGCA..12
T.a. TTTTCGAGTATATTGACAATTCCAAAAAACTGCTCATACTATGATTATAGTATAATCACG..32
O.s. TCTTCGAGTATATTGACAATTCCAAAAAACTGCTCATACTATCATTATAGTATAATGAGG..32
M.p. CTAAAAAAACTATTGACAGGAAATCATATAATTGAGTAAGATAAGGTTAAGGATAAA.....0
C.p. ATTCTAAAGTAGTTGACATATTTATTTTCATAGTA.TAATATTAAATATATAAATATATC...4
Cons ------A-tataTTGACAatt-caaaaaactg-tcaTActAT-A--atagta--at----
                "-35"                      "-10"
```

Figure 2 : Alignment of the 5'-flanking region of the cp *trn*E genes from various organisms. "-35" and "-10" indicate the conserved elements resembling the consensus *E. coli* promoter elements. The number at the right indicates the number of bases up to the beginning of the coding region. "Cons" indicate the consensus sequence : bases which are conserved in all cases are in capitals ; those conserved in all examined angiosperms are in lower case. The first 7 lines show the sequence flanking the chloroplast *trn*E gene from *Nicotiana tabacum*, *Vicia faba*, *Pisum sativum*, *Spinacia oleracea*, *Triticum aestivum*, *Oryza sativa* and *Marchantia polymorpha*, respectively. The corresponding region of the cyanelle genome from *Cyanophora paradoxa* (c.p.) is also shown.

However, transcription of some spinach cp tRNA genes was not abolished in this *in vitro* system when the 5' region of these genes was deleted (Gruissem et al. 1986), suggesting that other types of promoters, possibly internal to the coding region, are also involved in the transcription, at least for a sub-set of cp tRNA genes. This hypothesis is strengthened by the observation that many cp tRNA genes do not exhibit the typical "-35" and "-10" motif found upstream of *trn*E and of other chloroplast tRNA genes.

IV. Processing of chloroplast tRNAs

Cp tRNAs, like other RNAs, are synthesized as immature mono- or poly-cistronic precursors. In prokaryotes and eukaryotes, the 5'-leader sequence is removed by a site-specific endonuclease (RNase P) which cleaves precisely at the 5'-end of the mature tRNA . A RNase P-like activity has also been identified in the cp soluble transcription extract of *Euglena gracilis* and spinach chloroplasts (Greenberg et al. 1984a ; Greenberg and Hallick 1986) and in S30 lysates of spinach and tobacco chloroplasts (Yamaguchi-Shinozaki et al. 1987).
A RNase activity responsible for the 3'-terminus processing has been detected in the soluble transcription extracts (Greenberg et al. 1984a ; Greenberg and Hallick 1986) of spinach and *Euglena gracilis* chloroplasts and in S 30 lysates of spinach and tobacco chloroplasts (Yamaguchi-Shinozaki et al. 1987).

V. Concentrations of isoaccepting tRNAs and codon usage in chloroplasts

It has been shown in *E. coli* and in *Saccharomyces cerevisiae* that differences in the population of isoaccepting tRNAs are related to differences in the usage of synonymous codons and that there is a strong correlation between codon usage and tRNA content. It also has been shown in *E. coli* that the rate of translation is limited at some codons by the availability of the matching tRNAs and therefore that codon usage determines translation rate (Sorensen et al. 1989).

In order to see whether there is a correlation between codon usage and tRNA abundance in chloroplasts, we have compared on the one hand the codon usage in chloroplast protein genes from maize, spinach and tobacco and on the other the tRNA content in the chloroplast of bean (Pfitzinger et al., 1987). The amounts of bean cp tRNAs which can be charged with each of the 20 amino acids have been measured by aminoacylation of total cp tRNAs. We clearly showed that, in chloroplasts, there is a strong positive correlation between the relative abundance of the codons for each amino acid in cp protein genes and the relative abundance of the corresponding tRNAs. For example, codons for leucine are the most abundant, while the codon for tryptophan is the least abundant in cp protein genes and the amounts of $tRNA^{Leu}$ are the highest, while the amounts of $tRNA^{Trp}$ are the lowest.

Table I : Synonymous codon usage and relative content of isoacepting species in chloroplasts (modified from Pfitzinger et al. 1987).

tRNA corresponding to :	Arg	Arg	Gly	Gly	Ile	Ile	Leu	Leu	Leu (d)
ANTICODONS (b)	UCU	ACG	GCC	UCC	CAU (c)	GAU	UAA	CAA	UAG
CODONS	AGA AGG	CGU CGC CGA CGG	GGU GGC	GGA GGG	AUA	AUU AUC	UUA	UUG	CUU CUC CUA CUG
Codons usage in chloroplasts % (a)	31	69	49	51	26	74	36	21	43
Percentage of each isoacceptor in its tRNA family (in bean chloroplasts)	26	74	50	50	32	68	36	22	42

a) Codon usage in the chloroplasts of tobacco, maize and spinach was taken into account. The values represent the percentage of each codon in the family of synonymous codons.

b) Possible nucleotide modifications have not been considered.

c) Modification of C allows recognition of the isoleucine codon AUA and not of the methionine codon AUG (Francis and Dudock 1982).

d) Codon reading is acording to Pfitzinger et al. (1987).

We also examined for four amino acids (Arg, Gly, Ileu and Leu), whether a correlation exits between the frequency of codon usage in cp protein genes and the amounts of the corresponding isoacceptors. It appears that for these amino acids, the tRNA pool is adapted to the frequency of codon usage in the corresponding family. There is also a good correlation between the usage of synonymous codons and the relative concentrations of the corresponding isoaccepting tRNAs in chloroplasts (Table I). It is therefore likely that this adjustment reflects the selection for an optimum translation rate in chloroplasts.

The tRNA species encoded by genes present in duplicate on the cp genomes (in the IR sequences), are not found in higher amounts than the tRNAs encoded by single copy genes (Pfitzinger et al. 1987; Pillay et al. 1984), indicating that in chloroplasts, the relative concentrations of the various tRNAs are not controlled by gene dosage. It should also be noted that the presence of long introns does not seem to interfere with the expression of these tRNA genes, since all these tRNAs are found in relative high amounts.

VI. Regulation of plastid tRNA levels

It is likely that the strength of their respective promoters influence the relative levels of tRNAs in chloroplasts, as was demonstrated *in vitro* for the spinach *trnM*2 gene and derivatives of it, in which the genuine *trnM*2 promoter elements have been replaced by promoter elements from other cp genes (Gruissem and Zurawski 1985b).

Concerning $tRNA^{Phe}$, we have shown recently that the two isoacceptors species present in bean chloroplasts differ only by post-transcriptionnal modifications and that their relative amounts vary during leaf development : in etiolated leaves, the undermodified $tRNA^{Phe}$ only represents 15% of total cp $tRNA^{Phe}$, during development and greening it increases to reach 60% in 8-day-old leaves and then drops down to 9% in senescing leaves (Pfitzinger et al. 1990b).

Thus, although limited information is available, it appears likely that both transcriptional and post-transcriptional mechanisms of regulation contribute to determine the relative levels of tRNAs in chloroplasts.

The mechanisms which control the overall levels of cp tRNAs during development remain largely speculative, although it is known that cp tRNAs are present not only in photosynthetically active tissues but also (in lower amounts) in non-photosynthetic tissues such as roots, young embryogenic cotyledons or in dark-grown plants (Burkard et al. 1972 ; Merrick and Dure 1972). More recently, we have observed the presence of cp tRNAs in yellow senescent leaves, although the corresponding genes appeared not to be transcribed any longer in these organs (Jayabaskaran et al. 1989), whereas transcription of cp tRNA genes was detected in dark-grown maize plants (Jayabaskaran et al. unpublished results).

VII. Codon recognition in chloroplasts

As mentioned above, there are only 30 different tRNA species encoded by the higher plant cp genomes, whereas all 61 sense codons are found in cp protein genes and whereas, according to the "wobble" hypothesis, a minimum of 32 tRNA species would be required to decode the 61 codewords. Furthermore, since $tRNA^{Leu}$(UAA) and $tRNA^{Leu}$(CAA) which are present in chloroplasts should be able to decode the same leucine codon UUG, it appears that there is an apparent deficit of 3 tRNA species in higher plant chloroplasts. In fact, a comparison of the anticodons of the higher plant cp tRNAs with all sense codons shows that there is an apparent deficit of tRNAs which can normally recognize the following codons: GCU/C (Ala), CGC/A/G (Arg), CUU/C (Leu) and CCU/C (Pro) (Sugiura and Wakasugi 1989). Since it is generally assumed that there is no import of cp tRNAs (unlike some mitochondrial tRNAs which are imported from the cytosol), these observations suggest that a special mechanism of codon recognition must operate in chloroplasts, at least in some cases.

A "two-out-of-three" mechanism (only two bases of the anticodon get paired with two bases of the codon) has been shown to operate in an *in vitro* protein-synthezising system from *E. coli* (Samuelson et al. 1980). By *in vitro* translation experiments of rabbit globin mRNA or GAPDH mRNA using a tRNA-dependent wheat germ protein-synthesizing system that we have recently developed (Pfitzinger et al., 1989), we have shown that the single cp $tRNA^{Ala}$(U*GC), the single cp $tRNA^{Pro}$(U*GG) and the cp $tRNA^{Arg}$(ICG) can read all four alanine (GCN), proline (CCN) and arginine codons (CGN) respectively (Pfitzinger et al. 1990a). In the presence of these three cp tRNAs, a "two out of three" mechanism involving two strong G:C base pairs in the codon-anticodon interaction is operating. Concerning cp $tRNA^{Leu}$(UAm^7G), we also demonstrated by *in vitro* translation of synthetic polynucleotides that a "U:U wobble" and a "U:C wobble" mechanisms are occuring in order to read the CUU/C codons. In fact, in this case, the m^7G present at position 36 of the anticodon seems to facilitate the U:U and U:C wobble interaction. Thus, using a "two out of three" mechanism as well as a "U:N wobble" mechanism, the 30 chloroplast-encoded tRNAs should be able to read all 61 sense codons. It is also interesting to note that a comparison of total and polysome-bound tRNAs has shown that all higher plant cp tRNA species are apparently used in protein biosynthesis (Pfitzinger et al 1987).

VIII. Peculiar features of chloroplast $tRNA^{Ile}$(C*AU)

In plant chloroplasts, two $tRNAs^{Ile}$ are found ; the major $tRNA^{Ile}$ (GAU) recognizes the codons AUU and AUC, while the minor $tRNA^{Ile}$ (C*AU) probably recognizes the codon AUA. Surprisingly, the gene coding for the minor spinach chloroplast $tRNA^{Ile}$ was found to contain a CAU anticodon which is normally a methionine anticodon (Kashdan and Dudock,

1982). In this case, the C residue is modified post-transcriptionally into C* and Francis and Dudock (1982) have demonstrated that this $tRNA^{Ile}$(C*AU) is aminoacylated with isoleucine but not with methionine. Similar observations have been made in the case of bacteriophage T4 (Guthrie and McClain, 1979) and *E. coli* $tRNA^{Ile}$. In *E. coli*, the anticodon of the minor $tRNA^{Ile}$ gene is CAU, but this $tRNA^{Ile}$ has a lysidine (L) at the first position of its anticodon which allows it to recognize the AUA codon (Muramatsu et al, 1988a and 1988b). We also demonstrated recently that, in potato mitochondria, a $tRNA^{Ile}$(L*AU) containing a derivative of lysidine (L*) at position 34 is coded for by a mitochondrial gene possessing a methionine anticodon (Weber et al, 1990).

IX. Protein synthesis in organelles requires misaminoacylation of $tRNA^{Gln}$

Three glutamate accepting tRNA species could be identified in barley chloroplasts (Schön et al., 1988). The first $tRNA^{Glu}$ (Schön et al., 1986) is involved in the biosynthesis of -aminolevulinate (a precursor of chlorophyll) and probably in the incorporation of glutamate into proteins. The two other glutamate accepting tRNA species (differing only by a post-transcriptional modification) have both a U*UG anticodon specific for glutamine. These mischarged Glu-$tRNA^{Gln}$ species can be converted into Gln-$tRNA^{Gln}$ by a crude chloroplast extract (Schön et al., 1988). This reaction requires a specific amido-transferase and glutamine or asparagine as amide donors.

REFERENCES

Burkard, G. , Vaultier, J. P. , and Weil, J. H. (1972). Differences in the level of plastid-specific tRNAs in chloroplasts and etioplasts of *Phaseolus vulgaris*. Phytochem. 11,1351-1353

El-Gewely, M. R. , Helling, R. B. , and Dibbits, J. G. T. (1984). Sequence and evolution of the regions between the *rrn* operons in the chloroplast genome of *Euglena gracilis bacillaris*. Mol. Gen. Genet. 194,432-443

Evrard, J. L. , Kuntz, M. , Straus, N. A. , and Weil, J. H. (1988). A class-I intron in a cyanelle tRNA gene from *Cyanophora paradoxa*: phylogenetic relationship between cyanelles and plant chloroplasts. Gene 71,115-122

Francis, M. , and Dudock, B. (1982). Nucleotide sequence of a spinach chloroplast isoleucine tRNA . J. Biol. Chem. 257,11195-11198

Greenberg, B. M. , Gruissem, W. , and Hallick, R. B. (1984a). Accurate processing and pseudouridylation of chloroplast transfer RNA in a chloroplast transcription system. Plant Mol. Biol. 3,97-109

Greenberg, B. M. , and Hallick, R. B. (1986). Accurate transcription and processing of 19 *Euglena* chloroplast tRNAs in a *Euglena* soluble extract. Plant Mol. Biol. 6,89-100

Gruissem, W. , and Zurawski, G. (1985a). Identification and mutational analysis of the promoter for a spinach chloroplast transfer RNA gene. EMBO J. 4,1637-1644.

Gruissem, W. , and Zurawski, G. (1985b). Analysis of promoter regions for the spinach chloroplast *rbcL, atpB* and *psbA* genes. EMBO J. 4,3375-3383

Gruissem, W. , Elsner-Menzel, C. , Latshaw, S. , Narita, J. O. , Schaffer, M. A. , and Zurawski, G. (1986). A subpopulation of spinach chloroplast tRNA genes does not require upstream promoter elements for transcription. Nucl. Acids Res. 14,7541-7556

Guthrie, C. , and McClain, W. H. (1979). Rare transfer ribonucleic acid essential for phage growth. Nucleotide sequence comparison of normal and mutant T4 isoleucine-accepting transfer ribonucleic acid. Biochem. 18,3786-3795

Hiratsuka, J. , Shimada, H. , Whittier, R. , Ishibashi, T. , Sakamoto, M. , Mori, M. , Kondo, C. , Honji, Y. , Sun, C. R. , Meng, B. Y. , Li, Y. Q. , Kanno, A. , Nishizawa, Y. , Hirai, A. , Shinozaki, K. , and Sugiura, M. (1989). The complete sequence of the rice (*Oryza sativa*) chloroplast genome: Intermolecular recombination between distinct tRNA genes accounts for a major plastid DNA inversion during the evolution of the cereals. Mol. Gen. Genet. 217,185-194

Howe, C. J. (1985). The endpoints of an inversion in wheat chloroplast DNA are associated with short repeated sequences containing homology to att-lambda. Curr. Genet. 10,139-145

Jayabaskaran, C. , Kuntz, M. , Guillemaut, P. , and Weil, J. H. (1989). Variations in the levels of chloroplast tRNAs and aminoacyl-tRNA synthetases in senescing leaves of

Phaseolus vulgaris. Plant Physiol. 92, 136-140

Kashdan, M. A. , and Dudock, B. S. (1982). The gene for a spinach chloroplast isoleucine tRNA has a methionine anticodon. J. Biol. Chem. 257,11191-11194

Markowicz, Y. , Mache, R. , and Loiseaux-de Goër, S. (1988). Sequence of the plastid rDNA spacer of the brown alga *Pylaiella littoralis* (L.) Kjellm. Evolutionary significance. Plant Mol. Biol. 10,465-469

Merrick, W. C. , and Dure, L. S. (1972). The developmental biochemistry of cotton seed embryogenesis and germination. IV. Levels of cytoplasmic and chloroplastic transfer ribonucleic acid species. J. Biol. Chem. 247,7988-7999

Michel, F. , and Dujon, B. (1983). Conservation of RNA secondary structures in two intron families including mitochondrial-, chloroplast-, and nuclear-encoded members. EMBO J. 2,33-38

Muramatsu, T. , Nishihawa, K. , Nemoto, F. , Kuchino, Y. , Nishinura, S. , Miyazawa, T. , and Yokoyama, S. (1988a). Codon and amino acid specificities of a transfer RNA are both converted by a single post-transcriptional modification. Nature 336,179-181.

Muramatsu, T. , Yokoyama, S. , Horie, N. , Matsuda, A. , Veda, T. , Yamaizumi, Z. , Kuchino, Y. , Nishinura, S. , and Miyazawa, T. (1988b). A novel lysine-substituted nucleoside in the first position of the anticodon of minor isoleucine tRNA from *Escherichia coli*. J. Biol. Chem. 263,9261-9267

Ohme, M. , Kamogashira, T. , Shinozaki, K. , and Sugiura, M. (1985). Structure and cotranscription of tobacco chloroplast genes for $tRNA^{Glu}$(UCC), $tRNA^{Tyr}$(GUA) and $tRNA^{Asp}$(GUC). Nucl. Acids Res. 13,1045-1056

Ohyama, K. , Fukuzawa, H. , Kohchi, T. , Shirai, H. , Sano, T. , Sano, S. , Umesono, K. , Shiki, Y. , Takeuchi, M. , Chang, Z. , Aota, S. , Inokuchi, H. , and Ozeki, H. (1986). Chloroplast gene organization deduced from complete sequence of liverwort *Marchantia polymorpha* chloroplast DNA. Nature 322,572-574

Oliver, R. P. and Poulsen, C. (1984). Structure of a heavily transcribed region of barley chloroplast DNA. Transfer RNA genes for serine (UGA), glycine (GCC,UCC), formyl-methionine and threonine (GGU). Carlberg Res. Commun. 49,647-673

Pfitzinger, H., Guillemaut, P., Weil, J. H., and Pillay, D. T. N. (1987). Adjustment of the tRNA population to the codon usage in chloroplasts. Nucl. Acids Res. 15,1377-1386

Pfitzinger, H., Weil, J.H., Pillay, D.T.N. and Guillemaut, P. (1989) Preparation of a tRNA-dependent wheat germ protein synthesizing system. Plant Mol. Biol. 12, 301-306

Pfitzinger, H., Weil, J.H., Pillay, D.T.N. and Guillemaut, P. (1990a). Codon recognition mechanisms in plant chloroplasts. Plant Mol. Biol. 14, 805-814

Pfitzinger, H., Maréchal-Drouard L., Pillay, D.T.N., Weil, J.H. and Guillemaut, P.

(1990b) Variations during leaf development of the relative amounts of two bean (*Phaseolus vulgaris*) chloroplast tRNAs^{Phe} which differ in their minor nucleotide content. Plant Mol. Biol. 14, 969-975

Pillay, D. T. N. , Singh, G. P. , and Guillemaut, P. (1984). Localization of four genes for three tRNA^{Leu}s on the physical map of the soybean (*Glycine max* L.) chloroplast genome. Gene 29,351-354

Quigley, F. , and Weil, J. H. (1985). Organization and sequence of five tRNA genes and of an unidentified reading frame in the wheat chloroplast genome: evidence for gene rearrangement during the evolution of chloroplast genomes. Curr. Genet. 9,495-503.

Samuelson, T. , Elias, P. , Lusting, F. , Axberg, T. , Fölsch, G. , Akeson, B. , and Lagerkvist, U. (1980). Aberrations of the classic codon reading scheme during protein synthesis in vitro. J. Biol. Chem. 255,4583-4588

Schön, A. , Krupp, G. , Gough, S. , Berry-Lowe, S. , Kannangara, C. G. , and Söll, D. (1986). The RNA required in the first step of chlorophyll biosynthesis is a chloroplast glutamate tRNA. Nature 322,281-284

Schön, A. , Kannangara, C. G. , Gough, S. , and Söll, D. (1988). Protein biosynthesis in organelles requires misaminoacylation of tRNA. Nature 331,187-190

Shinozaki, K. , Ohme, M. , Tanaka, M. , Wakasugi, T. , Hayashida, N. , Matsubayashi, T. , Zaita, N. , Chunwongse, J. , Obokata, J. , Yamaguchi-Shinozaki, K. , Ohto, C. , Torazawa, K. , Meng, B. Y. , Sugita, M. , Deno, H. , Kamogashira, T. , Yamada, K. , Kusuda, J. , Takaiwa, F. , Kato, A., Tohdoh, N. , Shimida, H. , and Sugiura, M. (1986). The complete nucleotide sequence of the tobacco chloroplast genome: its gene organization and expression. EMBO J. 5,2043-2049

Sorensen, M. A. , Kurland, C. G. , and Pedersen, S. (1989). Codon usage determines translation rate in *Escherichia coli*. J. Mol. Biol. 207,365-377

Sugiura, M. , and Wakasugi, T. (1989). Compilation and comparison of transfer RNA genes from tobacco chloroplasts. Critical Rev. Plant Sci. 8,89-101

Weber, F., Dietrich, A., Weil, J.H. and Maréchal-Drouard L. (1990) A potato mitochondrial isoleucine tRNA is coded for by a mitochondrial gene possessing a methionine anticodon. Nucleic Acids Res., in press

Yamaguchi-Shinozaki, K. , Shinozaki, K. , and Sugiura, M. (1987). Processing of precursor tRNAs in a chloroplast lysate. FEBS Lett. 215,132-136

CHLOROPLAST GENES CODING FOR RIBOSOMAL PROTEINS IN LAND PLANTS

Masahiro Sugiura, Keita Torazawa and Tatsuya Wakasugi

Center for Gene Research
Nagoya University
Nagoya 464-01
Japan

Introduction

Chloroplast ribosomes in land plants are 70S in size similar to prokaryotic ribosomes and contain 23S, 16S, 5S and 4.5S rRNAs and about 60 ribosomal proteins (for reviews see Palmer 1985, Sugiura 1989). All the rRNAs are known to be encoded in the chloroplast genome. Analyses of the synthesis of ribosomal proteins in isolated chloroplasts have shown that the chloroplast genome encodes about one-third of the ribosomal proteins (Eneas-Filho *et al* 1981, Dorne *et al* 1984b, Posno *et al* 1984). The rest of chloroplast ribosomal proteins are thought to be encoded in the nuclear genome. The electrophoretic patterns and molecular weight frequency distributions of the proteins of higher plant chloroplast ribosomes are quite similar to those of the *Escherichia coli* ribosomes (Mache *et al* 1980, Capel & Bourque *et al* 1981). Determination of the primary structure of ribosomal protein L12 from spinach chloroplasts (Bartsch *et al* 1982) and immunological studies (Dorne *et al* 1984a, Bartsch 1985) showed also the similarity of chloroplast and *E. coli* ribosomal proteins. These observations led us to search for chloroplast ribosomal protein genes through their homology with *E. coli* counterparts. The tobacco *rps19* sequence was thus found as the first potential gene for a ribosomal protein in the chloroplast genome (Sugita & Sugiura 1983), and subsequently additional putative chloroplast genes for ribosomal proteins have been reported in

NATO ASI Series, Vol. H 55
The Translational Apparatus of Photosynthetic Organelles
Edited by R. Mache et al.

a variety plants (e.g. Subramanian et al 1983).

After determination of the complete nucleotide sequences of tobacco, liverwort and rice chloroplast DNAs, twenty different reading frames potentially coding for proteins homologous to E. coli ribosomal proteins were found in each genome. Recently an additional gene for a ribosomal protein was found in tobacco chloroplast DNA (Yokoi et al 1990) and hence the above sequenced chloroplast genomes contain 21 different genes potentially coding for ribosomal proteins. Here we present a compilation of the structure of these 21 chloroplast ribosomal protein genes from the three sequenced genomes and the summary of transcription pattern of the tobacco 21 genes.

Chloroplast genes for ribosomal proteins

Sequence analysis of the chloroplast genome from tobacco, rice and liverwort revealed the presence of 21 different reading frames (the presence of introns was assumed in four cases) potentially coding for polypeptides homologous to E. coli ribosomal proteins (Shinozaki et al 1986b, Hiratsuka et al 1989, Ohyama et al 1986). rpl32 is a newly identified gene through determination of the N-terminal amino acid sequence of a tobacco ribosomal protein (Yokoi et al 1990). Table 1 summarizes these 21 genes (rpl for large subunit proteins and rps for small subunit proteins) and their deduced products (CL and CS for chloroplast large and small subunit proteins, respectively). The tobacco and rice genomes (and other angiosperms analyzed) contain rps16 but lack rpl21, so far rpl21 has been unique in liverwort. Legume chloroplast DNAs lack rpl22 (Spielmann et al 1988; Palmer et al 1988). At present we cannot exclude the possibility that there are additional chloroplast genes for ribosomal proteins which have little homology with any of the E. coli ribosomal proteins.

A map of the tobacco genes for ribosomal proteins is shown in Fig. 1. Among them 3'-rps12, rps7, rpl23 and rpl2 are located in the inverted repeat (IR) in tobacco. The rice IR has expanded outward to include rps19 and rps15 which are present in the single-copy region of tobacco. The liverwort IR is short and contains no genes for ribosomal proteins. The rpl23, rpl2, rps19, rpl22, rps3, rpl16, rpl14, rps8, rpl36 and rps11 sequences are clustered in this order and this arrangement corresponds to that of the homologous genes in the E. coli S10, spc and α operons (Tanaka et al 1986). This finding raises the interesting possibility that the genes for ribosomal proteins of chloroplast and E. coli may have evolved from a common ancestral gene set.

Homology scores with E. coli counterparts are distributed from 74 % (CS12) to 18 % (CL32). Homology of ribosomal proteins among the three plants (50 - 92 %) is lower than that of

Table 1. Chloroplast genes for ribosomal proteins

Gene	Intron (bp)			Number of residues[a)]				Homology[b)] (%)		
	T	R	L	T	R	L	E	T	R	L
rps2				236	236	235	241	41.9	37.2	44.4
rps3				218	239	217	233	34.8	27.9	39.5
rps4				201	201	202	206	37.4	38.3	39.3
rps7				155	156	155	178[c)]	39.3	37.6	39.3
rps8				134	136	132	130	40.8	42.3	46.9
rps11				138	143	130	129	56.6	54.3	50.4
rps12	536	540	500	123	124	123	124	74.2	67.7	70.2
rps14				100	103	100	99	43.4	43.4	46.5
rps15				87	90	88	89	34.8	39.3	33.7
rps16	860	809		85	62		82	37.8	25.6	
rps18				101	163	75	75	36.0	24.0	40.0
rps19				92	93	92	92	56.5	46.7	64.1
rpl2	666	663	544	274	273	277	273	49.1	48.4	52.4
rpl14				123	123	122	123	54.5	52.8	58.5
rpl16	1020	1020	535	134	136	143	136	55.1	53.7	53.1
rpl20				128	119	116	118	44.1	44.1	45.8
rpl21						116	103			33.0
rpl22				155	103	119	110	40.0	30.0	41.8
rpl23				93	93	91	100	23.0	27.0	26.0
rpl32				55	63	69	56	25.0	17.9	23.2
rpl33				66	66	65	55	45.5	45.5	49.1
rpl36				37	37	37	38	60.5	65.8	60.5

T, tobacco; R, rice; L, liverwort; E, E.coli

a) fMet is included.

b) Numbers of identical residues are divided by numbers of E.coli residues, Gaps are introduced to maximize homology.

c) S7K from E.coli K12.

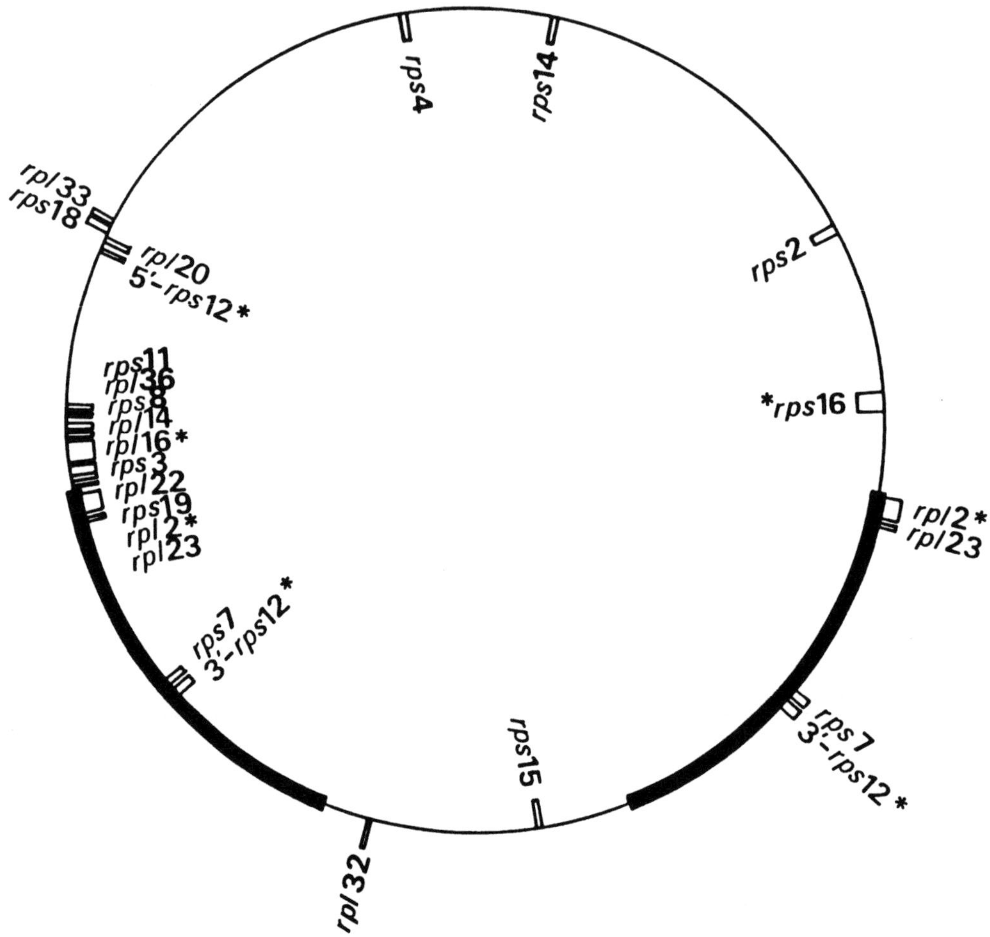

Fig. 1. Location of ribosomal protein genes on the tobacco chloroplast genome. Bold lines show the inverted repeat. Asterisks indicate split genes. Genes shown inside the circle are transcribed clockwise and genes shown outside the circle are transcribed counterclockwise.

components of the photosynthetic apparatus (71 - 100 %) (Shimada & Sugiura 1991). The deduced CL23 protein is low in homology with the E. coli L23 (23 - 27 %) and the rpl23 homologue in spinach and several other species is frame-shifted (Zurawski & Clegg 1987). In spinach, no gene products were detected from the split rpl23 sequence while transcripts were observed to initiate in front and within the rpl23 sequence and hence it would be a transcribed pseudogene (Thomas et al 1988). We found that the N-terminal amino acid sequence of one of the tobacco chloroplast ribosomal proteins matches that deduced from the tobacco rpl23 gene (Yokoi et al 1991). This indicates that rpl23 is functional in tobacco and probably in rice and liverwort. In rice and wheat, internally deleted rpl23 pseudogenes are present in the large single-copy region (LSC), an intact copy of the gene is still present in its normal position of the IR (Moon et al 1988; Ogihara et al 1988; Hiratsuka et al 1989). In addition, the rice genome contains two sequences homologous to portions of the 3'-rps12 intron in the LSC (Shimada & Sugiura 1989). These pseudogenes are not present in tobacco and liverwort. It remains a puzzle why apparently defective copies of these genes should accumulate in chloroplasts, which for the most part maintain compact genomes.

Most of the derived ribosomal proteins are similar in size among the three plants and E. coli except rps18 and rpl22. The rice CS18 (163 residues) is more than twice as long as the E. coli S18 (75 residues), which arises from a C-terminal extension. The size of CL22s varies from species to species, the tobacco CL22 of 155 residues has a C-terminal extension with respect to the E. coli L22 while both N-terminal and C-terminal extensions are present in the spinach CL22 of 199 residues (Zhou et al 1989). In contrast, the size of photosynthesis components among the three plants are highly conserved (Shimada & Sugiura 1991). Therefore the primary structures of chloroplast ribosomal proteins seem to be flexible for their functions.

Codon usage in the sum of 21 potential ribosomal protein genes is roughly similar to each other among the three plants

Table 2. Codon usages in chloroplast genes for ribosomal proteins

	U				C				A				G				
		T	R	L		T	R	L		T	R	L		T	R	L	
U	Phe	1.60	1.91	2.61	Ser	1.63	1.37	2.42	Tyr	2.65	2.30	2.75	Cys	1.05	1.15	1.06	U
	Phe	0.73	1.26	0.11	Ser	1.20	1.15	0.33	Tyr	0.36	0.58	0.29	Cys	0.11	0.25	0.22	C
	Leu	2.87	2.84	5.61	Ser	1.20	1.01	1.50	term	0.58	0.58	0.77	term	0.04	0.14	0.00	A
	Leu	1.74	1.83	0.55	Ser	0.58	0.54	0.15	term	0.15	0.04	0.00	Trp	0.94	0.79	0.92	G
C	Leu	1.63	1.51	1.50	Pro	1.27	1.40	1.61	His	2.03	2.01	1.54	Arg	2.90	3.02	2.83	U
	Leu	0.40	0.58	0.11	Pro	0.98	1.04	0.22	His	0.36	0.36	0.48	Arg	0.47	0.54	0.40	C
	Leu	0.83	0.72	0.40	Pro	1.02	1.01	1.80	Gln	2.76	2.81	3.82	Arg	2.90	2.27	2.28	A
	Leu	0.40	0.40	0.11	Pro	0.65	0.54	0.18	Gln	0.62	0.79	0.22	Arg	0.69	0.86	0.33	G
A	Ile	4.43	4.06	6.24	Thr	1.81	1.83	2.06	Asn	3.30	3.38	5.36	Ser	1.34	1.58	1.25	U
	Ile	1.45	1.40	0.29	Thr	1.20	0.94	0.40	Asn	1.16	1.44	0.51	Ser	0.29	0.29	0.18	C
	Ile	3.05	2.81	3.49	Thr	1.92	1.87	2.72	Lys	6.64	7.30	10.86	Arg	3.52	3.96	3.01	A
	Met	2.54	2.30	2.35	Thr	0.62	0.61	0.40	Lys	2.10	2.52	0.29	Arg	1.09	0.83	0.15	G
G	Val	1.85	2.19	2.50	Ala	2.25	2.09	2.57	Asp	2.32	1.94	2.39	Gly	2.50	2.27	2.17	U
	Val	0.87	0.58	0.26	Ala	1.02	1.19	0.51	Asp	0.83	0.79	0.26	Gly	0.44	0.65	0.29	C
	Val	2.10	2.12	1.83	Ala	2.54	2.27	2.35	Glu	3.37	3.38	3.89	Gly	2.94	2.66	3.27	A
	Val	0.65	0.61	0.11	Ala	0.40	0.72	0.18	Glu	0.98	0.79	0.26	Gly	1.12	1.04	0.48	G

The sum of 21 ribosomal protein genes.
T, tobacco; R, rice; L, liverwort.

(Table 2). All possible codons are used in the ribosomal protein genes except that only UAA is used as the termination codon in liverwort. Codons containing A and U are used more frequently than those containing G and C when compared within codons used for one amino acid, especially in liverwort.

An intron within a potential chloroplast ribosomal protein gene was first found in *Nicotiana debneyi rpl2* (Zurawski *et al* 1984). The *rps12*, *rps16*, *rpl2* and *rpl16* sequences contain introns (Fromm *et al* 1986, Shinozaki *et al* 1986ab, Sugiura *et al* 1987, Tanaka *et al* 1986, Ohyama *et al* 1986, Hiratsuka *et al* 1989). The most striking feature is that *rps12* consists of three exons and its 5' exon (*5'-rps12*) is located far away from the other two exons (*3'-rps12*). We have designated this gene structure a "divided" gene. Reverse transcription and electron microscopic analyses indicated that *trans*-splicing between tobacco *5'-rps12* and *3'-rps12* transcripts occurs *in vivo*. (Zaita *et al* 1987, Koller *et al* 1987, Hildebrand *et al* 1988).

Transcription of ribosomal protein genes from tobacco

Transcriptional analysis of the chloroplast genes for ribosomal proteins have been performed in varieties of higher plants (e.g. Ohto *et al* 1988, Thomas *et al* 1988, Neuhaus *et al* 1989). Northern blot hybridization was first carried out to detect transcripts from each gene. Transcription units were then determined using primer extension and S1 mapping analyses. *In vitro* capping experiments are necessary to determine the initiation sites of transcription.

Transcripts from all the 21 ribosomal protein genes were detected in tobacco chloroplasts (Ohto *et al* 1988, Yokoi *et al* 1990). *rps16* may be transcribed monocistronically (Shinozaki *et al* 1986a) while all the other *rps*/*rpl* genes are cotranscribed with their adjacent genes and show complex transcription patterns. The 21 genes fall into ten transcription units as shown in Table 3. The largest transcription unit is the *rpl23* cluster which includes 10 ribosomal protein genes

Table 3. Transcription units of ribosomal protein genes in tobacco

units	Transcripts largest[a)]	Transcripts major[b)]
rpl23-rpl2-rps19-rpl22-rps3-rpl16-rpl14-rps8-rpl36-rps11-(rpoA)	4.5	
(IRF196)-5'rps12-rpl20	3.6	1.2
(ORF31-petG-psaJ)-rpl33-rps18	4.2	2.8
rps4[c)]	4.6	1.3
(psaA-psaB)-rps14	5.2	5.2
(rpoB-rpoC1-rpoC2)-rps2	4.8	2.2
rps16	1.3	1.3, 0.4
3'rps12-rps7	2.4	1.2, 0.9
rps15-(ORF393-ndhA)	2.5	0.5
rpl32[c)]	1.5	1.5

( ) No ribosomal protein genes.
a) The largest transcripts detected and possibly primary transcripts (kb).
b) Major transcripts as estimated by band densities (kb).
c) Transcription unit remain unanalyzed.

and rpoA at its end.

Two separate transcripts from 5'-rps12 and 3'-rps12 are spliced in *trans* to produce a mature mRNA for CS12. Some of the ribosomal protein genes are cotranscribed with genes not functionally related, e.g. the psaA operon. A 5.2 kb transcript from the psaA-psaB-rps14 operon encodes two photosystem I components and CS14, the former is thought to be light-induced and the latter constitutive, strongly suggesting the presence of a post-transcriptional regulation system.

Acknowledgments

We thank Dr. M. Sugita, Dr. T. Matsubayashi and Mr. F. Yokoi for their suggestions and discussions. This work was supported by a Grant-in-Aid from the Ministry of Education, Science and Culture (Japan).

References

Bartsch M (1985) Correlation of chloroplast and bacterial ribosomal proteins by cross-reactions of antibodies specific to purified Escherichia coli ribosomal proteins. J Biol Chem 260:237-241

Bartsch M, Kimura M, Subramanian AR (1982) Purification, primary structure, and homology relationships of a chloroplast ribosomal protein. Proc Natl Acad Sci USA 79:6871-6875

Capel MS, Bourque DP (1982) Characterization of Nicotiana tabacum chloroplast and cytoplasmic ribosomal proteins. J Biol Chem 257:7746-7755

Dorne AM, Eneas-Filho J, Heizmann P, Mache R (1984a) Comparison of ribosomal proteins of chloroplast from spinach and of E. coli. Mol Gen Genet 193:129-134

Dorne AM, Lescure AM, Mache R (1984b) Site of synthesis of spinach chloroplast ribosomal proteins and formation of incomplete ribosomal particles in isolated chloroplasts. Plant Mol Biol 3:83-90

Eneas-Filho J, Hartley MR, Mache R (1981) Pea chloroplast ribosomal proteins: characterization and site of synthesis. Mol Gen Genet 184:484-488

Fromm H, Edelman M, Koller B, Goloubinoff P, Galun E (1986) The enigma of the gene coding for ribosomal protein S12 in the chloroplasts of Nicotiana. Nucl Acids Res 14:883-898

Hildebrand M, Hallick RB, Passavant CW, Bourque DP (1986) Trans-splicing in chloroplasts: The rps12 loci of Nicotiana tabacum. Proc Natl Acad Sci USA 85:372-376

Hiratsuka J, Shimada H, Whittier RF, Ishibashi T, Sakamoto M, Mori M, Kondo C, Honji Y, Sun CR, Meng BY, Li YQ, Kanno A, Nishizawa Y, Hirai A, Shinozaki K, Sugiura M (1989) The complete sequence of the rice (Oryza sativa) chloroplast genome: Intermolecular recombination between distinct tRNA genes accounts for a major plastid DNA inversion during the evolution of the cereals. Mol Gen Genet 217:185-194.

Koller B, Fromm H, Galun E, Edelman M (1987) Evidence for in vivo trans splicing of pre-mRNAs in tobacco chloroplasts. Cell 48:111-119

Mache R, Dorne AM, Batlle RM (1980) Characterization of spinach plastid ribosomal proteins by two dimensional gel electrophoresis. Mol Gen Genet 177:333-338

Meng BY, Tanaka M, Wakasugi T, Ohme M, Shinozaki K, Sugiura M (1988) Cotranscription of the genes for two P700 chlorophyll a apoproteins with the gene for ribosomal protein CS14: Determination of the transcriptional initiation site by in vitro capping. Curr Genet 14:395-400

Moon E, Kao TH, Wu R (1988) Rice mitochondrial genome contains a rearranged chloroplast gene cluster. Mol Gen Genet 213:247-253

Neuhaus H, Scholz A, Link G (1989) Structure and expression of a split chloroplast gene from mustard (Sinapis alba): ribosomal protein gene rps16 reveals unusual transcriptional features and complex RNA maturation. Curr Genet 15:63-70

Ogihara Y, Terachi T, Sasakuma T (1988) Intermolecular recombination of chloroplast genome mediated by short direct-repeat sequences in wheat species. Proc Natl Acad Sci USA 85:8573-8577

Ohto C, Torazawa K, Tanaka M, Shinozaki K, Sugiura M (1988) Transcription of ten ribosomal protein genes from tobacco chloroplast genome. Plant Mol Biol 11:589-600

Ohyama K, Fukuzawa H, Kohchi T, Shirai H, Sano T, Sano S, Umesono K, Shiki Y, Takeuchi M, Chang Z, Aota S, Inokuchi H, Ozeki H (1986) Chloroplast gene organization deduced from complete sequence of liverwort _Marchantia polymorpha_ chloroplast DNA. Nature 322:572-574

Palmer JD (1985) Comparative organization of chloroplast genomes. Annu Rev Genet 19:325-354

Palmer JD, Osorio B, Thompson WF (1988) Evolutionary significance of inversions in legume chloroplast DNAs. Curr Genet 14:65-74

Posno M, van Noort M, Debise R, Groot GSP (1984)Isolation, characterization, phosphorylation and site of synthesis of _Spinacia_ chloroplast ribosomal proteins. Curr Genet 8:147-154

Shimada H, Sugiura M (1989) Pseudogenes and short repeated sequences in the rice chloroplast genome. Curr Genet 16:293-301

Shimada H, Sugiura M (1991) Fine structural features of the chloroplast genome: comparison of the sequenced chloroplast genomes. to be submitted

Shinozaki K, Deno H, Sugita M, Kuramitsu S, Sugiura M (1986a) Intron in the gene for the ribosomal protein S16 of tobacco chloroplast and its conserved boundary sequences. Mol Gen Genet 202:1-5

Shinozaki K, Ohme M, Tanaka M, Wakasugi T, Hayashida N, Matsubayashi T, Zaita N, Chunwongse J, Obokata J, Shinozaki KY, Ohto C, Torazawa K, Meng BY, Sugita M, Deno H, Kamogashira T, Yamada K, Kusuda J, Takaiwa F, Kato A, Tohdoh N, Shimada H, Sugiura M (1986) The complete nucleotide sequence of the tobacco chloroplast genome: its gene organization and expression. EMBO J 5:2043-2049

Spielmann A, Roux E, von Allmen JM, Stutz E (1988) The soybean chloroplast genome: complete sequence of the _rps19_ gene, including flanking parts containing exon 2 of _rpl2_ (upstream) but lacking _rpl22_ (downstream). Nucl Acids Res 16:1199

Subramanian AR, Steinmetz A, Bogorad L (1983) Maize chloroplast DNA encodes a protein sequence homologous to the bacterial ribosome assembly protein S4. Nucl Acids Res 11:5277-5286

Sugita M, Sugiura M (1983) A putative gene of tobacco chloroplast coding for ribosomal protein similar to _E. coli_ ribosomal protein S19. Nucl Acids Res 11:1913-1918

Sugiura M, Shinozaki K, Tanaka M, Hayashida N, Wakasugi T, Matsubayashi T, Ohto C, Torazawa K, Meng BY, Hidaka T, Zaita N (1987) Split genes and _cis/trans_ splicing in tobacco chloroplasts. In: von Wettstein D, Chua NH (eds) Plant Molecular Biology. Plenum, New York pp 65-76

Tanaka M, Wakasugi T, Sugita M, Shinozaki K, Sugiura M (1986) Genes for the eight ribosomal proteins are clustered on the chloroplast genome of tobacco (Nicotiana tabacum): Similarity to the S10 and spc operons of Escherichia coli. Proc Natl Acad Sci USA 83:6030-6034

Thomas F, Massenet O, Dorne AM, Briat JF, Mache R (1988) Expression of the rpl23, rpl2 and rps19 genes in spinach chloroplasts. Nucl Acids Res 16:2461-2472

Yokoi F, Vassileva A, Hayashida N, Torazawa K, Wakasugi T, Sugiura M (1990) Chloroplast ribosomal protein L32 is encoded in the chloroplast genome. FEBS Lett submitted

Yokoi F, Tanaka M, Wakasugi T, Sugiura M (1991) The chloroplast gene for ribosomal protein CL23 is functional in tobacco. FEBS Lett to be submitted

Zaita N, Torazawa K, Shinozaki K, Sugiura M (1987) Trans splicing in vivo: joining of transcripts from the 'divided' gene for ribosomal protein S12 in the chloroplasts of tobacco. FEBS Lett 210:153-156

Zhou DX, Quigley F, Massenet O, Mache R (1989) Cotranscription of the S10- and spc-like operons in spinach chloroplasts and identification of three of their gene products. Mol Gen Genet 216:439-445

Zurawski G, Bottomley W, Whitfeld PR (1984) Junctions of the large single copy region and the inverted repeats in Spinacia oleracea and Nicotiana debneyi chloroplast DNA: sequence of the genes for tRNA-His and the ribosomal proteins S19 and L2. Nucl Acids Res 12:6547-6558

Zurawski G, Clegg MT (1987) Evolution of higher-plant chloroplast DNA-encoded genes: implications for structure-function and phylogenetic studies. Ann Rev Plant Physiol 38:391-418

THE EVOLUTION OF GENES AND PSEUDOGENES FOR SOME CHLOROPLAST RIBOSOMAL PROTEINS. TRANSPOSITION AND RECOMBINATION LEAD TO DIFFERENT FATES IN DIFFERENT GENOMES.

C.M.Bowman,
IPSR, Cambridge Laboratory,
Centre for Plant Science Research,
Colney Lane, Norwich NR4 7UJ, UK.

Transposition events have created chloroplast-DNA-derived ribosomal protein pseudogenes in chloroplast (Bowman et al 1988, Shimada and Sugiura 1989) and mitochondrial (eg Moon et al 1988) genomes of some higher plants. Examining the role of homologous recombination in the evolutionary fate of these ribosomal protein genes and their pseudogenes, can offer some insight into the possible influence of homologous recombination on the evolution of organelle genomes in general.

When natural crosses occur between plants with biparental plastid inheritance, there is no evidence for genetic recombination between parental chloroplast genomes. This clearly limits the ways in which homologous recombination can contribute to the evolution of plastid DNA, when compared with nuclear (see Dover and Tautz, 1986), or mitochondrial DNA. However, although there is a natural barrier to recombination between genomes from different plastids, there are many lines of evidence to suggest that homologous recombination probably (and perhaps readily) occurs within and between the many genome copies within a plastid. Most compelling is the extensive homologous recombination revealed in plants regenerated from intraspecific (Medgyesy et al 1985) or interspecific (Thanh and Medgyesy 1989) protoplast fusions. The dimers (Kolodner and Tewari 1979) and oligomers (Deng et al 1989) seen in ctDNA preparations are also thought to be products of intermolecular homologous recombination. A last and well-known example is that of

NATO ASI Series, Vol. H 55
The Translational Apparatus of Photosynthetic Organelles
Edited by R. Mache et al.

intramolecular recombination between the two segments of the ctDNA large inverted repeat (Palmer 1983).

Thus, if homologous recombination is detected when plastid genomes carrying appropriate markers are given access to one another, it is equally probable that it occurs within the "essentially" homogeneous population of genomes within a plastid. This being so, how might it contribute to the evolution of those genomes ? In the nucleus it is believed that homologous recombination between high copy-number (eg rDNA) genes contributes to a process termed molecular drive, and has a powerful "driving " influence on the evolution of those genes (Dover and Tautz 1896). Does homologous recombination make an equivalent contribution to the evolution of multicopy organelle genomes ?

During studies on ribosomal protein genes, it has been possible to detect products of recombination between some of the ctDNA ribosomal protein genes and their pseudogene homologues. In the following paper, I briefly consider the consequences of these recombination events on the ribosomal protein gene homologues themselves, and by implication, on the evolution of the genomes that contain them.

Ribosomal protein genes and their pseudogene homologues

Pseudogene sequences have been recognised by DNA/DNA hybridisation and sequencing of cloned wheat and maize ctDNA, and maize mtDNA (Bowman and Dyer 1986, Bowman et al 1988, Moon et al 1988), and also by computer analysis of the complete rice ctDNA sequence (Shimada and Sugiura 1989). The particular ribosomal protein pseudogene sequences discussed in this paper are homologous with the gene sequences rpl2, rpl23 and rps12, all of which are located in the large inverted repeat region of the chloroplast genome as illustrated in Fig.1.

Before the divergence of wheat, maize and rice, the pseudogenes were created by transposition events that inserted copies or segments of these genes elsewhere in the ctDNA, as illustrated in Fig.2. (Since the recognisable pseudogene sequences

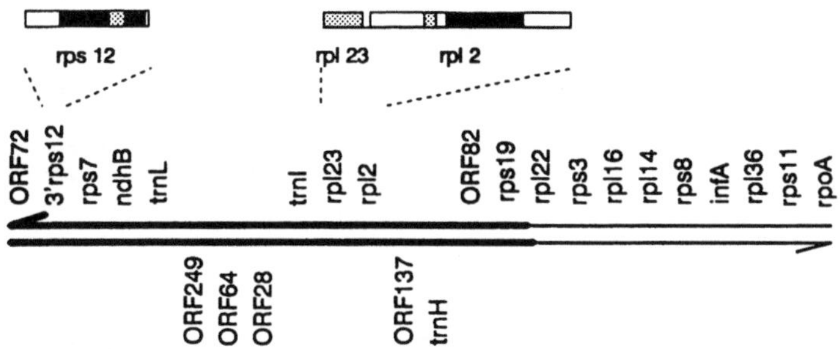

Fig.1. Chloroplast ribosomal protein genes containing homology with pseudogenes. A region of the circular genome is shown with the gene order indicated on both DNA strands. Thick lines represent inverted repeat DNA. The organisation of rps12, rpl23 and rpl2 is detailed above. The segments duplicated elsewhere to form pseudogenes, are stippled. Introns are blackened.

reflect the sequence conserved rather than the sequence transposed, nucleotides flanking the duplications cannot be used to diagnose the transposition mechanisms). Subsequently, large segments of ctDNA (designated "promiscuous" DNA, Stern and Lonsdale 1982, Ellis 1982) were transferred to the mtDNA, carrying with them the ctDNA rpl2-rpl23 sequences, and the ctDNA rbcL-ψrpl23-ORF133 sequences (see Stern and Lonsdale 1982, Moon et al 1988).

Thus, as a result of "early" transposition events, gene and pseudogene homologues of rpl2 and rpl23 have been evolving independently in the chloroplast and mitochondrial genomes of several monocot species. Some of these genomes are diagrammed in Fig.2 [The spinach ctDNA rpl23 carries a deletion which implies that in spinach, rpl23 is a nuclear gene (Zurawski and Clegg 1987). Also, the entire ψrpl23 sequence has apparently been deleted from one of the wild wheats (Aegilops spp., Bowman and Dyer 1986)]. Examination of the available sequences provides evidence for recombination between rpl23 homologues in both types

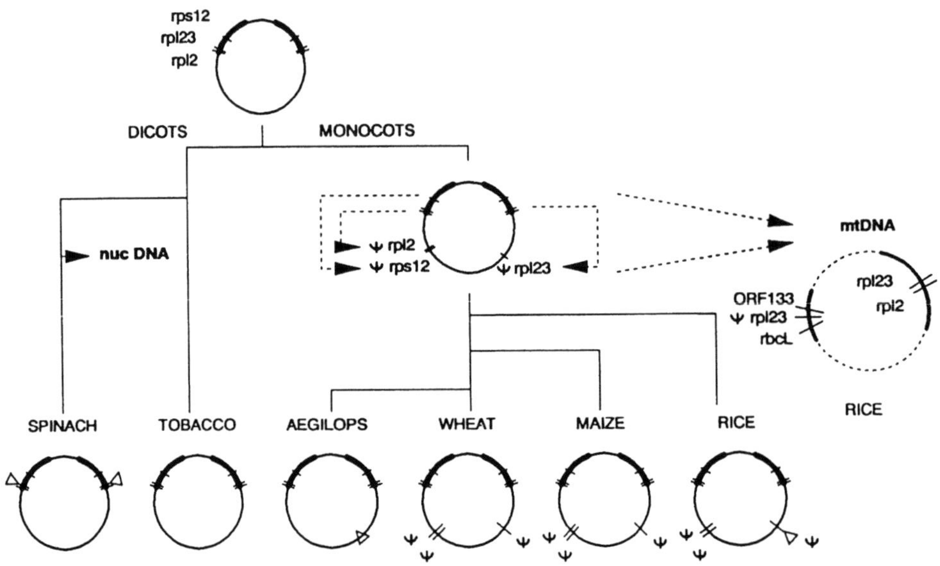

Fig.2. Transposition and divergence of some rpl23, rpl2 and rps12 homologues in higher plant organelle genomes. In ctDNA: thick lines represent inverted repeat DNA, | = gene location, ∇ = deletion in gene, -∇- = deletion of gene. In mtDNA: thick lines represent "ctDNA".

of organelle genome. In chloroplasts, circumstantial evidence comes from conservation of the ψrpl23 homologues, while in mitochondria, direct evidence comes from crossing over between the marker genes that flank the rpl23 homologues.

Conservation of rpl23 pseudogenes in chloroplast DNA

The extreme similarity between the wheat (Bowman et al 1988), maize (McLaughlin and Larrinua 1988), rice (Hiratsuka et al 1989) and tobacco (Shinozaki et al 1986) rpl23 sequences is demonstrated by their alignment in Fig.3a. The wheat and maize ψrpl23 sequences (Bowman et al 1988) are also extremely similar (90% homology, Fig.3b). In the rice ψrpl23 sequence however

(a)

```
               .         .         .         .         .         .         .
T.a. ATGGATGGAATCAAATACGCAGTATTTACAGAAAAAAGTCTTCGTTTATTGGGAAAGAATCAATATACTTTTAATGTCGA  80
Z.m.                                    G
O.s.
N.t.                  T              C      A    G        G  A             C

     .         .         .         .         .         .         .         .
T.a. ATCGGGATTCACTAAGACAGAAATAAAGCATTGGGTCGAACTCTTCTTTGGTGTTAAGGTGGTAGCTGTGAATAGCCATC  160
Z.m.                                                             A
O.s.                                                             A
N.t.    A    CA    G                                       C     AA     A       T

     .         .         .         .         .         .         .         .
T.a. GACTACCTGGAAAAGGTAGAAGAATAGGACCTATTCTGGGCCATACAATGCATTACAGACGTATGATCATTACCCTTCAA  240
Z.m.        C     G           G              A
O.s.        C     G           G  C        A  A
N.t.     T  G     GA          G         A    A                                G

     .         .         .         .         .
T.a. CCGGGTTATTCTATTCCACTTCTAGATAGAGAAAAAAACTAA  282
Z.m.                                   CG
O.s.
N.t.                    C   .T G A      G    T
```

(b)

```
               .         .         .         .         .         .         .
T.a. TCTATATATCTAGAATATATATATTAA.GGTATAATACAATAATGAAATACAAA.TAAAATAATAAAATATAGTATTATC  80
Z.m.                    G ....  C               AC     T   C .......C
O.s.          T     G    ....                   AT      ... ..........      G
       ______________________________________________________________________________
     .         .         .         .         .         .         .         .
T.a. ATATGATAATTGAATCAAATACGCAGTATTTACTGAAATAAGTCTTCGTTTATTGGGTTTGAATCAATATACTTTTAATG  160
Z.m. G  C      G                      A    AG                 AAA
O.s.           G                      A    .    T   A         AAA        ............
     ________________________________________________________________________________
     .         .         .         .         .         .         .         .
T.a. TCGAAT.GGGATTCACTAAGACTGAAATAAAGCATTGGGTCGTACTCTTCTTTGGTGTTAAGGTGGTAGCTGTGTATAGC  240
Z.m.       C               A                   A                     A         A
O.s. ...... .........................................................................
     ________________________________________________________________________________
     .         .         .         .         .         .         .         .
T.a. CATCGACTACCTAGAAA.GGTAAAAGAATAGGACCTATTCTGGGCCATACAATGCATTACAGACGTATGATCATTACCCT  320
Z.m.            CG    G    G      G              A
O.s. ................. ..........................A
     ______________________________________________
     .         .         .         .         .         .         .         .
T.a. TCAACCGGGTTATTCTATTCCACTTCTAGATAGAGAAAAAAACTAAAGGAGAATGAATGAAAAAAGACATAGTTTGGAAG  400
Z.m.      T   .                  CT.....TTT    T      GT T ..C     G G TAT TC   TATTT
O.s.  T                                                                     G

     .         .         .         .
T.a. TTAGACCTTTTTATAGGACTCTCTTTCAATTTCAAA  436
Z.m.  C C ATAGC TCTTTTGAA ...           G
O.s.       CC    C  A           AAAAG   G
```

Fig. 3. Comparison of rpl23 homologues. Nucleotide sequences (a) of rpl23 homologues, or (b) containing ψrpl23 homologues from wheat (T.a.), maize (Z.m.), rice (O.s.) and tobacco (N.t.), were aligned using UWGCG software. Dots represent gaps introduced to maximise homology. The wheat sequences are shown complete. Only nucleotides that differ are shown in the sequences aligned below. In (b) the ψrpl23 homology is overlined. (For sequence sources see text.)

(Shimada and Sugiura 1989), a 155bp deletion reduces the overall homology, while sequences flanking the deletion remain almost identical to the wheat pseudogene sequence. Alignment of the pseudogenes and their flanking sequences consequently reveals a degree of *ψrpl23* conservation similar to that of *rpl23* itself.

In the absence of data from divergent species, the extreme sequence homology between gene and pseudogene from the same species (see Bowman *et al*, 1988), might suggest a recent duplication. In these examples however, assuming a single event (see Fig.2), duplication of the gene is most likely to have occurred before the divergence of wheat, maize and rice. Sufficient time has elapsed to allow complete divergence between the wheat and maize sequences downstream of *ψrpl23*, and yet the wheat and maize pseudogene sequences are still 90% homologous. (In rice and wheat, the conservation of open reading frames immediately downstream of *ψrpl23*, complicates interpretation, see Shimada and Sugiura 1989). What kind of mechanism could constrain the divergence of the *ψrpl23* sequences? If enough divergence has accumulated between two homologues, the characteristics of that divergence can be used to recognise the kind of mechanism that could have allowed it to occur. Table 1 compares the base changes that broadly characterise divergence between two genes, with those that characterise divergence between a gene and its pseudogene. (With some exceptions, divergence between the wheat and tobacco *rpl23* homologues exemplifies these characteristics, for detailed data and discussion, see Bowman *et al* 1988).

Although the wheat and tobacco gene sequences have diverged by 11%, the amino acid sequence of the functional polypeptide is, as one would expect, conserved in a manner characteristic of a phenotypic selection mechanism. The wheat *rpl23* gene and pseudogene have diverged by only 5%, but in contrast, although the DNA sequence is highly conserved, the amino acid sequence is not. The data imply a mechanism conserving the DNA itself. An obvious candidate is some mechanism of gene conversion (see Szostak *et al* 1983, Kourilsky 1986). This can result from

Table.1. Characteristics of divergence between rpl23 homologues.

	GENE/GENE eg Wheat rpl23 x Tobacco rpl23	GENE/PSEUDOGENE eg Wheat rpl23 x Wheat ψrpl23
General distribution of base changes	More near 3' end of gene	Domains conserved irrespective of gene structure
Distribution between codon positions	Most at 3rd. position	Equally distributed between positions
Synonymous : nonsynonymous changes	Synonymous ≃ nonsynonymous	Most nonsynonymous (stop codons created)
Deletions/ Insertions	Absent/ rare, If frameshifts, corrected.	Present, therefore frameshifts
Therefore:	AMINO ACID SEQUENCE CONSERVED	DNA SEQUENCE CONSERVED
Implied conservation mechanism	Phenotypic selection for protein function	Mechanism operating at the level of the DNA

mismatch repair through homologous recombination.

The characteristics of ψrpl23 conservation thus provide circumstantial evidence for homologous recombination between rpl23 homologues in chloroplast DNA. If such a mechanism were operating, it predicts that crossing over might also occur (see Fig.4). The resulting recombinant plastid genomes, although they may exist transiently, would be unstable or lethal. The predicted recombination products are however detected in rice mitochondrial DNA (see below).

Crossing-over between rpl23 homologues in mitochondrial DNA

During studies on rice organellar genes, Moon et al (1988) recovered cloned mtDNA containing the ctDNA gene sequence rbcL-rpl23-rpl2. This is a predicted product from crossing over within the rpl23 homologues, and between the "parental" sequences rbcL-ψrpl23-ORF133 and trnI-rpl23-rpl2.

Figure 4 illustrates how, according to the Meselson and Radding model (see Szostak et al 1983, Kourilsky 1986), conservation of the chloroplast ψrpl23 homologues, and crossing over within the mitochondrial "ctDNA" rpl23 homologues, can result from alternative outcomes of equivalent recombination events in chloroplast and mitochondrion.

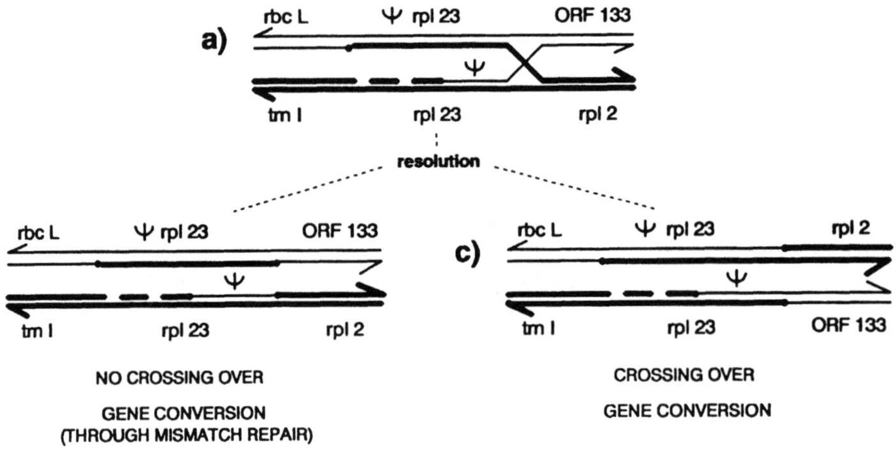

Fig.4. Alternative products of recombination between rpl23 and ψrpl23 homologues. Thick lines represent inverted repeat DNA (For explanation see text.)

After pairing of the rpl23 and ψrpl23 homologues, (from the single-copy and inverted repeat regions of the genome respectively), strand invasion, DNA replication and ligation occurs. Migration of the resulting Holliday junction forms heteroduplex DNA in which rpl23 and ψrpl23 complement one another (as shown in Fig.4 panel a). This complex can resolve to yield two types of recombination product, diagrammed in panels (b) and (c). In (b) crossing over has not occurred, and the patch of asymmetric heteroduplex is flanked by original parent duplex. This is the product inferred in ctDNA. Downstream of rbcL, mismatch repair can "convert" any diverged ψrpl23 sequence using the complementary rpl23 as template. It is assumed this happens frequently enough in ctDNA to conserve the ψrpl23 sequence and its ability to pair with rpl23. In the alternative outcome (c), crossing over has occurred and marker genes flanking the heteroduplex DNA have recombined. This is the type of product detected in mtDNA. Mismatch repair between rpl23 homologues is still possible.

Implications for Organelle Evolution

For convenience, genome evolution can be considered at two interdependent levels: that of structural rearrangement, and that of sequence divergence . Assuming that recombination between rpl23 homologues represents the behavior of the organelle genomes containing them, then homologous recombination can influence the evolution of organelle genomes at both levels. The rpl23 recombination products detected in chloroplast and mitochondrial DNA show that recombination between homologues can influence the divergence of those homologues (Fig. 4b), and/or lead to crossing over within them. The crossing over alternative results in structural rearrangement of the genome.

Homologous recombination may be an important component of several features that characterise plant organelle genomes and their evolution. These are summarised in Table 2. At face value, some features are shared, eg the relatively low rates of sequence divergence, while others are not, eg the differing degrees of structural rearrangement. This is to be expected because the primary influence of homologous recombination will be mediated differently in chloroplast and mitochondrion by factors such as

Table.2. Some features characteristic of plant organelle genomes, and their evolution.

CHLOROPLAST	MITOCHONDRIAL
Homologous recombination between inverted repeats generates isomers in equilibrium. Inverted repeat structure conserved[a].	Homologous recombination between direct repeats generates recombinant DNA circles in equilibrium[b].
Little other structural rearrangement[c].	i.e. Much structural rearrangement.
Inverted repeat sequence divergence is slow. (1/3 of rate in single-copy region[d])	Repeat sequence divergence is slow.
Rate of silent substitutions in gene sequences is low. (1/2 of rate in plant nuclear DNA[d])	Rate of silent substitution in gene sequences is low. (1/3 of rate in ctDNA[d]) (1/5 of rate in plant nuclear DNA[d]) (mammalian mtDNA 5x rate in mammalian nuclear DNA[d])

[a] Palmer 1983
[b] Lonsdale et al 1988
[c] Deng et al 1989
[d] Wolfe et al 1987

: organelle fusion, genome structure, recombination frequency, mutation and segregation rates, and selection pressure.
At the level of structural rearrangement, the mitochondrion appears to provide a relaxed selective environment. Genes are few, large repeats are many and recombination between them is very frequent. The resulting structural rearrangements characterise the organisation of mitochondrial genomes, in which recombinant circles exist in equilibrium (see Lonsdale et al 1988). In the absence of selection against such rearrangements, homologous recombination makes an important contribution to the

rapid evolution of mitochondrial genome structure. Every plant species so far examined has a unique mitochondrial gene order (see Palmer 1990).

By comparison, the selective environment of the chloroplast is a stringent one; the genome is tightly packed with genes, many of which are organised in transcription units (eg Shinozaki et al 1986). Frequent intramolecular homologous recombination between the two inverted repeat segments probably conserves the repeat DNA sequence by mismatch repair, while perpetuation of the characteristic inverted repeat structure is an almost inevitable consequence (Palmer 1983). In contrast, recombination between the few dispersed repeats, such as the ribosomal protein gene homologues, is probably less frequent and has fewer viable outcomes. Gene conversion, as inferred for the rpl23 homologues, is one of those viable alternatives. Crossing over will in many cases (including that of rpl23), create lethal structural variants that are unstable, delete genes, or interrupt transcription units (for discussion, see Bowman et al 1988). Inversions are more often viable, and are a relatively common evolutionary rearrangement. In complete contrast to the situation in plant mitochondria therefore, selection seems to predominate. Most structural rearrangement is eliminated and the evolutionary contribution of homologous recombination is minimised. With the notable exception of pea, chloroplast genome structure evolves very slowly. Most higher plant ctDNAs have an almost identical gene order (see Palmer 1990).

At the level of sequence divergence, homologous recombination may contribute to the relatively low rate of silent substitution observed in plant organelle genes (Table 2),(Wolfe et al 1987). The importance of its contribution will however depend on the frequency of recombination between the multicopy genomes within an organelle. Intermolecular gene conversion could accelerate the elimination of variant genomes, because there would be a high probability of a low-frequency variant being "converted" by a high-frequency wild type. This would reduce the rate of fixation of new mutants in organelle genome populations, and thus, the rate of silent substitution (for discussion see Bowman et al 1988). For two reasons, such a mechanism may have a greater influence on the evolution of plant mitochondrial rather than chloroplast genomes. The recombination frequency is probably higher, and fusion of plant mitochondria allows all the mitochondrial genomes within a cell to mix and recombine readily (see Lonsdale et al 1988).

Literature references

Bowman CM, Dyer TA (1986) The location and possible evolutionary signicance of small dispersed repeats in wheat ctDNA Curr Genet 10:931-941

Bowman CM, Barker RF, Dyer TA (1988) In wheat ctDNA, segments of ribosomal protein genes are dispersed repeats, probably conserved by nonreciprocal recombination. Curr Genet 14:127-136

Deng X-W, Wing RA, Gruissem W (1989) The chloroplast genome exists in multimeric forms. Proc Natl Acad Sci USA 86:4156-4160

Dover GA, Tautz G (1986) Conservation and divergence in multigene families: alternatives to selection and drift. Phil Trans R Soc Lond B 312:275-289

Ellis J (1982) Promiscuous DNA-chloroplast genes inside plant mitochondria. Nature 299:678-679

Hiratsuka J, Shimada H, Whittier R, Ishibashi T, Sakamoto M, Mori M, Kondo C, Honji Y, Sun C-R, Meng B-Y, Li Y-Q, Kanno A, Nishizawa Y, Hirai A, Shinozaki K, Sugiura M (1989) The complete sequence of the rice (Oryza sativa) chloroplast genome: Intermolecular recombination between distinct tRNA genes accounts for a major plastid DNA inversion during the evolution of the cereals. Mol Gen Genet 217:185-194

Kolodner R, Tewari KK (1979) Inverted repeats in chloroplast DNA from higher plants. Proc Natl Acad Sci USA 76:41-45

Kourilsky P (1986) Molecular mechanisms for gene conversion in higher cells. TIG 2:60-63

Lonsdale DM, Brears T, Hodge TP, Melville SE, Rottman WH (1988) The plant mitochondrial genome: homologous recombination as a mechanism for generating heterogeneity. In: Mitochondrial Biogenesis. Leaver CJ, Lonsdale DM (eds) Phil Trans R Soc Lond B pp149-163

McLaughlin WE, Larrinua, IM (1988) The sequence of the maize plastid encoded rpl23 locus. Nucl Acids Res 16:8183

Medgyesy P, Fejes E, Maliga P (1985) Interspecific chloroplast recombination in a Nicotiana somatic hybrid. Proc Natl Acad Sci USA 82:6960-6964

Moon E, Kao T-H, Wu R (1988) Rice mitochondrial genome contains a rearranged chloroplast gene cluster. Mol Gen Genet 213:247-253

Palmer JD (1983) Chloroplast DNA exists in two orientations.

Nature 301:92-93
Palmer JD (1990) Contrasting modes and tempos of genome evolution in land plant organelles. TIG 6:115-120
Shimada H, Sugiura M (1989) Pseudogenes and short repeated sequences in the rice chloroplast genome. Curr Genet 16:293-301
Shinozaki K, Ohme M, Tanaka M, Wakasugi T, Hayashida N, Matsubayashi T, Zaita N, Chunwongse J, Obokata J, Shinozaki KY, Ohto C, Torazawa K, Meng BY, Sugita M, Deno H, Kamogashira T, Yamada K, Kusuda J, Takaiwa F, Kato A, Tohdoh N, Shimada H, Sugiura M (1986) The complete nucleotide sequence of the tobacco chloroplast genome: its gene organisation and expression. EMBO J 5:2043-2049
Stern B, Lonsdale DM (1982) Mitochondrial and chloroplast genomes of maize have a 12-kilobase DNA sequence in common. Nature 299:698-702
Sjostak JW, Orr-Weaver TL, Rothstein RJ, Stahl FW (1983) The double-strand-break-repair model for recombination. Cell 33:25-35
Thanh ND, Medgyesy P (1989) Limited chloroplast gene transfer via recombination overcomes plastome-genome incompatibility between Nicotiana Tabacum and Solanum tuberosum. Plant Mol Biol 12:87-93
Wolfe KH, Li W-H, Sharp PM (1987) Rates of nucleotide substitution vary greatly among plant mitochondrial, chloroplast and nuclear DNAs. Proc Natl Acad Sci USA 84:9054-9058
Zurawski G, Clegg MT (1987) Evolution of higher-plant chloroplast DNA-encoded genes: implications for structure-function and phylogenetic studies. Ann Rev Plant Physiol 38:391-418

EXPRESSION OF NUCLEAR AND CHLOROPLAST GENES CODING FOR TOBACCO CHLOROPLAST RIBOSOMAL PROTEINS

Don P. Bourque[1,2], Gasmalla Elhag[1], Peta Bonham-Smith[3], Frank Thomas[4], Thomas McCreery[1], and Betty Glinsmann-Gibson[1]
Department of Biochemistry[1] and Department of Molecular and Cellular Biology[2], University of Arizona, Tucson AZ 85721 U.S.A.

INTRODUCTION

Coordinate expression of nuclear and chloroplast genes is necessary for the biogenesis of chloroplast ribosomes. In order to elucidate mechanisms which might regulate their expression, genes from both cellular compartments must be characterized. Genes coding for 20 chloroplast DNA-encoded ribosomal proteins of Nicotiana tabacum (tobacco) are known (reviewed in Bonham-Smith and Bourque, 1990). The remaining (over 35; Capel and Bourque, 1982) tobacco chloroplast ribosomal proteins must be coded by as yet uncharacterized nuclear genes. Here some features of deduced amino acid sequences and gene multiplicity for nuclear-encoded tobacco chloroplast ribosomal proteins L12, L24, and L27 are described. Comparisons with related E. coli and plant genes are presented.

Sequence analysis of products of gene expression (mRNA, polypetides) provides insights into gene structure and function, allowing predictions of specific mechanisms which might control their expression. Mechanisms which could operate at translational, and other posttranscriptional levels, to regulate chloroplast DNA-encoded protein expression have been proposed (Bonham-Smith and Bourque, 1989, 1990). Possible modulation of translational efficiency by ribosome recognition site locations and secondary structural features of chloroplast DNA-encoded mRNAs are discussed. In addition, possible autoregulatory feedback mechanisms for chloroplast ribosomal protein genes expression are examined.

[3]Current address: Department of Biological Sciences, University of Calgary, University Drive NW, Calgary, Alberta, Canada T2N 1N4
[4]Current address: Laboratoire de Physiologie Cellulaire Vegetale, Centre National de la Recherche Scientifique, Universite Scientifique et Medicale de Grenoble, F-38041 Grenoble-Cedex, France

NATO ASI Series, Vol. H 55
The Translational Apparatus of Photosynthetic Organelles
Edited by R. Mache et al.

CHLOROPLAST 50S SUBUNIT RIBOSOMAL PROTEINS L12, L24, AND L27 ARE CODED BY TOBACCO NUCLEAR GENES

From a cDNA library cloned in bacteriophage lambda gt11, full-length cDNAs for three nuclear-encoded tobacco chloroplast ribosomal proteins were identified and sequenced (Elhag and Bourque, unpublished results). Each these cDNAs encode a chloroplast ribosomal protein related to an E. coli ribosomal protein. The deduced protein sequences are homologous to E. coli L12 (involved in GTP hydrolysis), L24 (an rRNA binding protein important in assembly of 50S subunits), and L27 (a constituent of peptidyl transferase). Whether each of these predicted proteins is functionally equivalent to its procaryotic counterpart remains to be demonstrated experimentally. The cDNAs and predicted polypeptide sequences (Figures 1-3) show several common features (Table 1). Relative to the homologous E. coli sequences, the

TABLE 1: Features of cDNA and predicted polypeptides L12, L24, AND L27

50S Subunit Protein	cDNA Sequence (bp)	Residues in Mature Protein	Transit Peptide Length	Carboxyl-terminal Extension	Total Protein Length
L12	773	133	53	none	186
L24	792	127	70(?)	24	197
L27	882	128	51	40	179

deduced primary translation products each have an amino terminal (transit) peptide. Both L24 and L27 sequences have a carboxyl-terminal extension, a feature lacking in the L12 sequence. Using Northern blot analysis of total leaf and poly (A+) mRNA probed with cloned cDNA inserts, unique full-length cytoplasmic mRNAs coding for these proteins were observed which corresponded to the sizes of the largest cDNAs which were sequenced (Table 1). In each case (L12, L24, L27), at least two cDNAs with identical coding sequences and different 3'-noncoding sequences were observed. Furthermore, Southern blot analysis of genomic DNA restriction fragments revealed complex patterns of bands hybridizing to specific probes. These data and copy number reconstructions suggested that there are many genomic sequences coding for L12 and L27, but only a few genes coding for L24.

The predicted tobacco L12 sequence (Figure 1) has high homology with chloroplast ribosomal protein L12 of spinach (79%) and E. coli (46%). Using HPLC-purified protein, the N-terminal dodecapeptide (residues 58-69) of the mature protein was confirmed by protein microsequencing. The N-terminal sequence and the predicted transit peptide cleavage site (between amino acids 57 and 58) are in agreement with results of Bartsch, et al.(1982), Giese and Subramanian(1989), and Sugiura(these proceedings). Hydropathicity analysis predicts a highly conserved structure, and presumably function, between chloroplast and eubacterial L12 proteins. The amino acid composition of the putative transit peptide is similar to that of other chloroplast proteins (Schmidt and Mishkind, 1986).

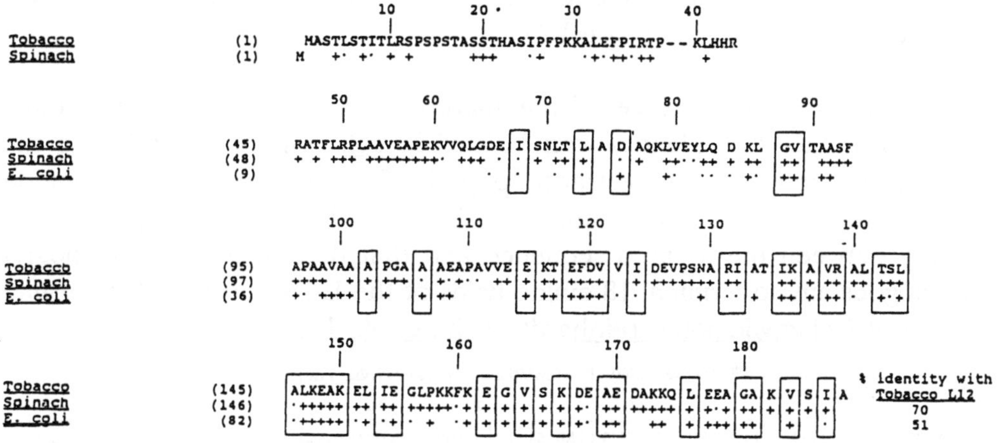

FIGURE 1: Amino acid sequence alignment of tobacco and spinach L12 chloroplast ribosomal proteins with that of E. coli.

Tobacco chloroplast ribosomal protein L24 has 76% identity with that of pea (Gantt, 1988) and is 35% identical with L24 of E. coli (Figure 2). The transit peptide is likely to be approximately 70 amino acids long, assuming the precursor would be processed as previously predicted by SDS gel analysis for pea L24 (Gantt, 1988). Highly conserved carboxyl-terminal extensions, relative to E. coli L24, occur in both the tobacco and pea L24 sequences. Further study is required to show whether these extensions have important functional properties, perhaps being involved in transport, ribosome assembly, or even carboxyl-terminal processing.

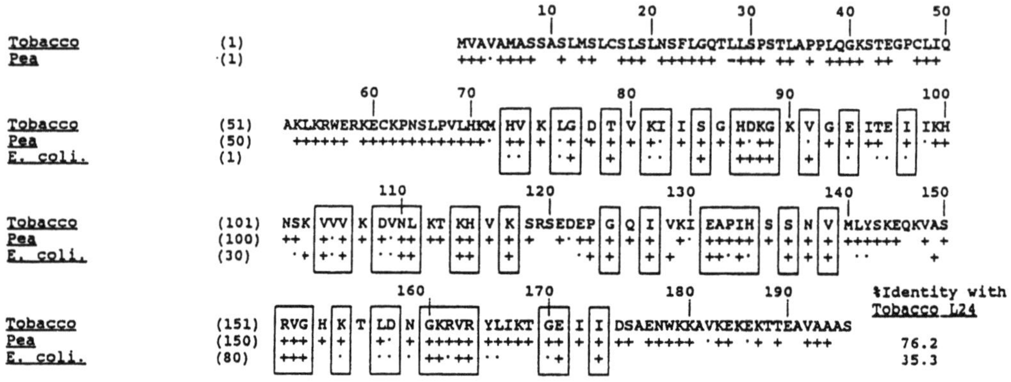

FIGURE 2: Amino acid sequence alignment of tobacco and pea chloroplast L24 proteins with that of E. coli L24.

The deduced sequence of the mature tobacco chloroplast ribosomal protein L27 has 78% identity with L27 of E. coli (Figure 3), excluding the putative 51 amino acid long transit peptide. The proposed transit peptide cleavage site (between Ala 51 and His 52) is based on the striking sequence identity between the tobacco sequence and that determined for the clearly homologous amino terminal 18 amino acids of the L-18 chloroplast ribosomal protein of Chlamydomonas reinhardtii (Liu, et al., 1988).

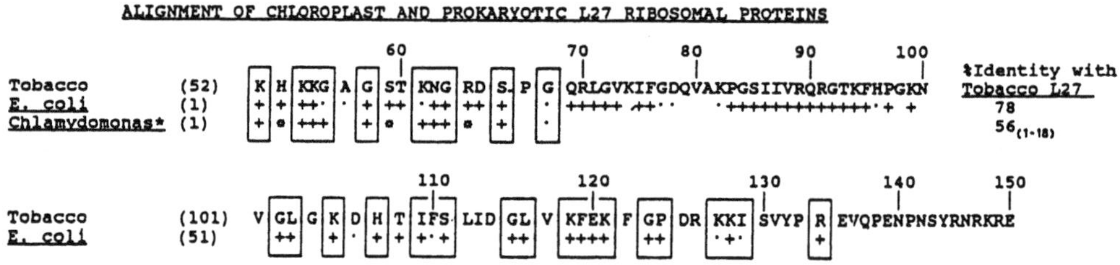

FIGURE 3: Amino acid sequence alignment of tobacco and Chlamydomonas L27-like proteins with that of E. coli L27.

TRANSLATIONAL REGULATION OF CHLOROPLAST mRNAs

Mechanisms which regulate translation of chloroplast-encoded genes are poorly understood. We analyzed sequences of 196 chloroplast polypeptide-coding genes of all protein classes to identify potential signals which might be involved in modulation of mRNA translational (Bonham-Smith and Bourque, 1989, 1990). Our surveys revealed a preference for AUG and UAA codons for initiation and termination of translation, respectively. Although most often observed within about 25 nucleotides upstream from the initiation codon(Figure 4A), a Shine-Dalgarno (SD) sequence (GGAGG or a tri- or tetranucleotide variant) was found within 100 bp 5' to the AUG codon in 92% of the genes. In 40% of these cases, the SD sequence location was similar to that of the consensus for procaryotes (-12 to -7, 5' to AUG). Results consistent with our were also obtained by Ruf and Kossel (1988).

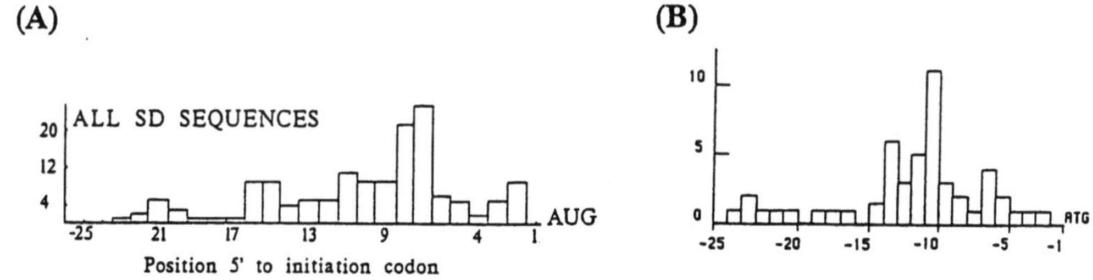

FIGURE 4: Location of Shine-Dalgarno sequences within 25 nucleotides upstream of initiation codons in chloroplast mRNA of (A) all protein classes and (B) ribosomal proteins.

Analysis of 81 chloroplast-encoded ribosomal protein genes also showed AUG and UAA to be the preferred codons for initiation and termination, respectively. A Shine-Dalgarno sequence was present in 84% of the sequences at a distance ranging from -2 to -100 bp, with a peak occurrence between positions -13 and -5.3) and the procaryotic consensus location occurred most frequently (Figure 4B). However, mRNA secondary structures may be involved in positioning more distal (-25 to -100 nucleotides) SD sequences to the presumably optimal location (-12 to -7) necessary to produce a stable translation initiation complex. For instance, in the case of chloroplast ribosomal protein genes (Figure 5), base-paired stem-loop structures can be drawn which bring short, distal SD sequence variants to within 7 bases upstream from the AUG codons. Comparative analysis of sequences at the 3'-end of chloroplast, mitochondrial, and procaryotic protein coding sequences (Bonham-Smith and Bourque, 1989) suggested that termination at UGA codons in chloroplast genes may occur by a mechanism which has been proposed for UGA termination in E. coli, and which involves 16S rRNA secondary structure. Whether termination rates might control translation of chloroplast mRNA remains to be tested experimentally.

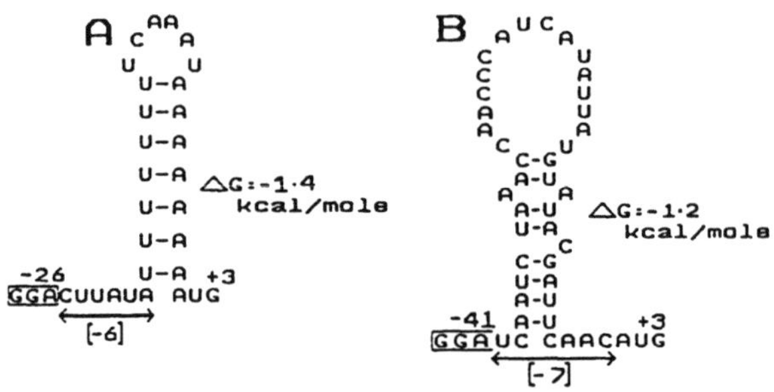

FIGURE 5: Proposed stem-loop structures within the leader sequence of (A) pea rps11 and (B) tobacco rps12 genes. The GGA Shine-Dalgarno sequence variants are boxed.

AUTOREGULATION OF CHLOROPLAST RIBOSOMAL PROTEIN GENE EXPRESSION

Chloroplast ribosomal protein expression could be regulated at several levels - transcription, mRNA processing and translation. Considering the numerous procaryotic properties of chloroplast genes, the simplest models to propose for regulatory mechanisms are well-established and include transcription regulatory proteins, such as the lactose and lambda repressors and/or translational regulation resulting from a translational feedback mechanism similar to those which occur in E. coli ribosomal protein operons (reviewed in Bonham-Smith and Bourque, 1990). The potential for feedback control of chloroplast ribosomal protein translation is interesting in that products of chloroplast as well as nuclear genes would be required, if the mechanisms are related to the bacterial autoregulatory feedback mechanism. The E. coli ribosomal protein operons code for regulatory proteins S8, S4, S7, S2 and S15, respectively (Lindahl and Zengel, 1986) for which a homologous polypeptide is coded by chloroplast DNA in tobacco and Marchantia. Other chloroplast ribosomal proteins (homologous to E. coli proteins L1, L10, L4, S1 and S20, which regulate translation of mRNA transcribed from the L1, L10, L4,

S1 and S20 operons, respectively), are likely to be products of plant nuclear genes. However, chloroplast ribosomal proteins which regulate translation may be homologous to those of E. coli operons, or different ribosomal or non-ribosomal proteins (chloroplast or nuclear-coded) may be translational regulators.

One common feature of E. coli ribosomal repressor proteins is their involvement in binding to rRNA during ribosome assembly. It is likely that rRNA and mRNA bind these regulatory proteins competitively with a higher affinity for rRNA than for mRNA. Therefore, if rRNA is in excess, these proteins would enter the ribosomal assembly pathway without inhibiting translation (Nomura, 1986). In support of this hypothesis is that the presumed mRNA binding sites for ribosomal proteins S4, S7 and S8 have sequences and secondary structural features which are similar to their respective binding sites on 16S rRNA.

Since tobacco and E. coli 16S rRNA genes share high sequence identity, we examined the possibility that both 16S rRNA and mRNA secondary structures to could be involved in feedback regulation of translation. In E. coli, ribosomal protein S7 inhibits it own translation by binding to the intercistronic

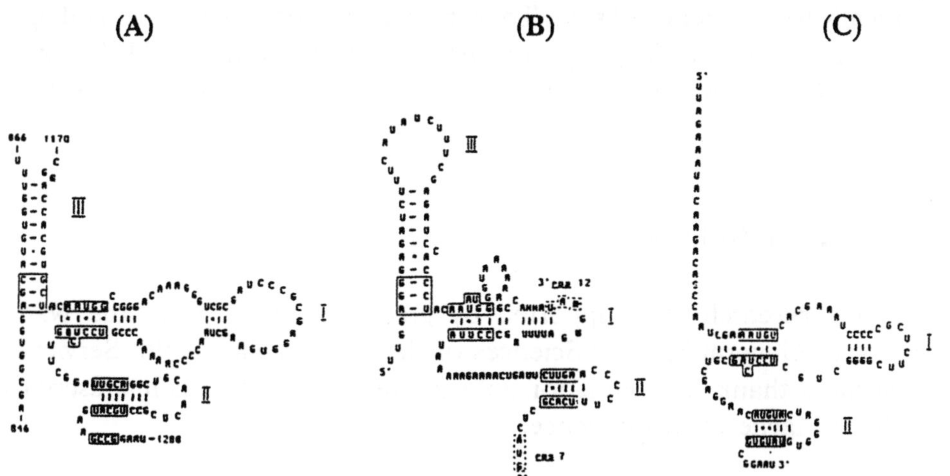

FIGURE 6: Predicted tobacco chloroplast ribosomal protein S7 binding sites on (A) 16S rRNA, (B) the rps12 (exon 3)-rps7 intercistronic sequence, and (C) the 3'-end of the transcript containing the rps12 exon 1 coding sequence .

region of the transcript carrying the rps12-rps7 coding sequences. In the tobacco chloroplast genome, ribosomal protein S12 is encoded on two separate transcripts that are trans-spliced to produce a dicistronic mRNA in which the S12 coding sequence is continuous (Hildebrand, et al.,1988; reviewed in Bonham-Smith and Bourque, 1990). Possible binding sites for chloroplast ribosomal protein S7 were identified on 16S rRNA, the intercistronic sequence (between rps12 exon 3 and rps7) and at the 3'-end of the exon 1 sequence (Figure 6[A-C], respectively). If ribosomal protein S7 does bind to both transcripts of the chloroplast rps12-rps7 operon, it could affect:

1) transcription (particularly of the rps12 exon 1 transcript),
2) the trans-splicing of exons 1 and 2 of rps12,
3) other mRNA maturation events such as splicing of the intron between exons 2 and 3 of rps12, or
4) translation of mRNAs for S12 and S7 by an autoregulatory feedback mechanism.

Binding sites on 16S rRNA and mRNA have also been identified for chloroplast ribosomal proteins S4 and S8 (Bonham-Smith and Bourque, unpublished results). Using cloned genes transcribed in vitro, we are testing whether tobacco chloroplast ribosomal protein S7, in particular, indeed binds specifically to the predicted sites. To test the hypothesis that these binding sites may be involved, in vitro experiments must show whether HPLC-purified chloroplast ribosomal proteins S7 affects expression in any of the suggested ways.

ACKNOWLEDGEMENTS

This research was supported by grant GM26937 from the National Institute of General Medical Sciences of the U.S. Public Health Service. The authors thank Jane Huff for help in preparation of this manuscript and Dr. Regis Mache for his patience.

REFERENCES

Bartsch, M, Kimura, M, Subramanian, AR (1982) Purification, primary structure and homology relationships of a chloroplast ribosomal protein. Proc Natl Acad Sci 79: 6871-6975.

Bonham-Smith, PC, Bourque, DP (1988) Translation of chloroplast-encoded mRNAs: potential initiation and termination signals. Nucleic Acids Res 17: 2057-2080.

Bonham-Smith, PC, Bourque, DP (1988) The chloroplast genome and regulation of its expression. In Adolph, KW (ed) Chromosomes: eucaryotic, procaryotic and viral, vol II. CRC Press, Boca Raton, pp 179-216.

Capel, MS, Bourque, DP (1982) Characterization of Nicotiana tabacum chloroplast and cytoplasmic ribosomal proteins. J Biol Chem 257: 7746-7755.

Gantt, JS (1988) Nucleotide sequences of cDNAs encoding four complete nuclear-encoded plastid ribosomal proteins. Curr Genet 14: 519-528.

Giese, K, Subramanian, AR (1989) Chloroplast ribosomal protein L12 is encoded in the nucleus: construction and identification of its cDNA clones and nucleotide sequence including the transit peptide. Biochem 28:3525-3529.

Hildebrand, M, Hallick, RB, Passavant, CW, Bourque, DP (1988) Trans-splicing in chloroplasts: the rps12 loci of Nicotiana tabacum. Proc Natl Acad Sci 85: 372-376.

Lindahl, L, Zengel,JM (1986) Ribosomal genes in Escherichia coli. Ann Rev Genet 20: 297-326.

Liu, X-Q, Gillham, NW, Boynton, JE (1988) Chloroplast ribosomal protein L-18 in Chlamydomonas reinhardtii is processed during ribosome assembly. Mol Gen Genet 214: 588-591.

Nomura, M (1986) Regulation of the synthesis of ribosomes and ribosomal components in Escherichia coli; translational regulation and feedback loops. In Booth, I, Loops, C (eds) Regulation of Gene Expression. Cambridge University Press, Cambridge, pp 199-220.

Ruf M, Kossel, H (1988) Occurrence and spacing of ribosome recognition sites in mRNAs of chloroplasts from higher plants. FEBS Letters 240: 41-44.

Schmidt, GM, Mishkind, ML (1986) The transport of proteins into chloroplasts. Ann Rev Biochem. 55: 879-912.

NUCLEAR-CODED CHLOROPLAST r-PROTEINS, PRECURSOR cDNA CLONES AND TRANSIT SEQUENCES

Alap R. Subramanian
Max-Planck-Institut für Molekulare Genetik
Abteilung Wittmann
Berlin-Dahlem, Germany

In this chapter I shall discuss some of the main findings from our recent work on the isolation and characterization of complementary DNA clones for six nuclear-coded proteins of the spinach chloroplast ribosome. The clones encode the precursor forms of these proteins which are synthesized in the cytoplasm. From protein-chemical data the chloroplast targeting sequences were identified and, in some cases, the amino acid sequences of the mature proteins were confirmed. The results show that nuclear-coded proteins have certain features which could not have been predicted from the data on the complementary organelle-encoded ribosomal proteins.

Chloroplast ribosomes of higher plants contain over 60 ribosomal proteins[1], of which 20-22 are encoded in the plastid DNA. That leaves roughly 40 proteins whose genes must be located in another cellular compartment. It is assumed that all the latter are located in the nuclear DNA, i.e. the plant mitochondrial DNA may not encode functional genes for chloroplast ribosomal proteins. Of these 40 genes, 14 have been characterized at the cDNA level at the time of this meeting.

We chose spinach (Spinacia oleracea) for our work because the strategy we were going to use required large amounts of chloroplast ribosomes, hence large amounts of an easily available leaf tissue material. Interestingly, the existence of

---

[1]For a correlation of the protein nomenclature used by different labs and 2-D gel patterns showing the protein spots identified so far, see a recent review article, Subramanian et al 1991. The properties of chloroplast-encoded r-proteins and the operon organization of their genes have been reviewed in Subramanian et al 1990.

NATO ASI Series, Vol. H 55
The Translational Apparatus of Photosynthetic Organelles
Edited by R. Mache et al.

chloroplast ribosomes was first shown in spinach (Lyttleton, 1962). Later Mache and coworkers used it for a detailed analysis of the protein pattern of chloroplast ribosomes (Mache et al 1980; Dorne et al 1984). The complete amino acid sequence of a chloroplast ribosomal protein, which provided the experimental support for homology assignment between ribosomal proteins of the chloroplast and _E. coli_, was done on a protein (L12) isolated from spinach (Bartsch et al 1982).

Cloning Strategy

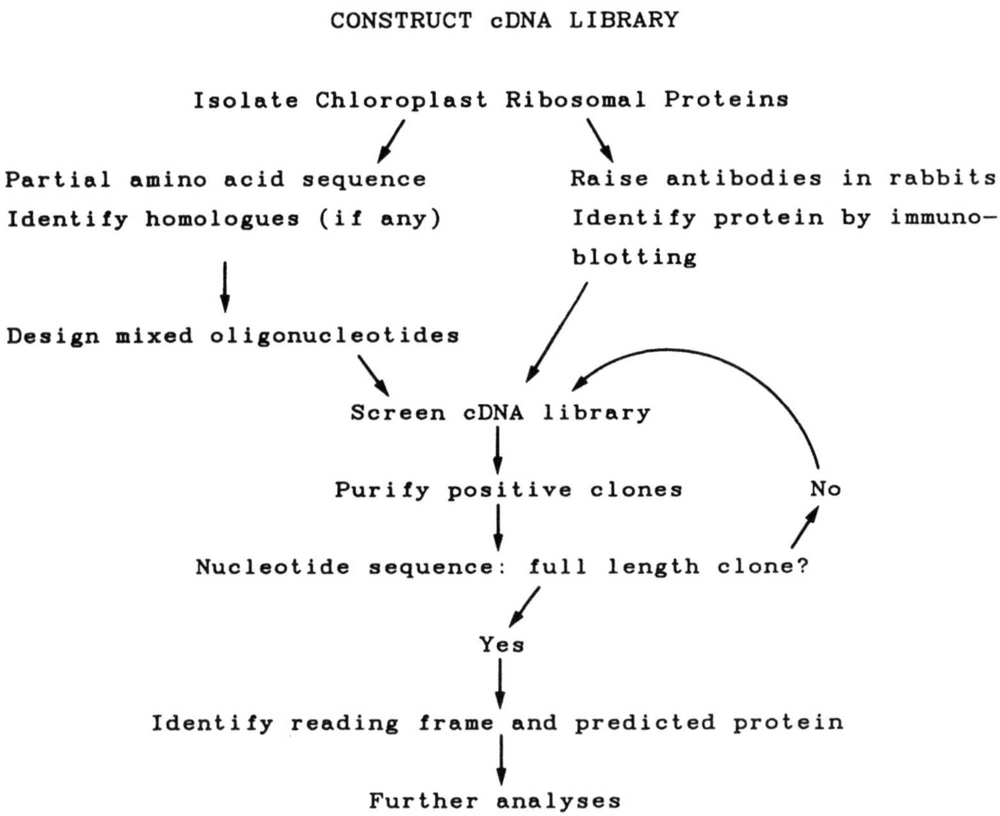

Fig. 1: Strategy for isolating cDNA clones for nuclear-coded chloroplast ribosomal proteins (written by Dr. P.M. Smooker of my group)

An outline of the cloning strategy we used is shown in Fig. 1. Two sets of cDNA libraries were constructed, in λgt10 and in the expression vector λgt11, as described elsewhere (Giese and Subramanian 1989; Phua et al 1989).

We took the precaution of methylating possible EcoR1 sites in the cDNA to block cleavage at these sites during the subsequent steps of library construction. As it turned out cDNAs for several ribosomal proteins (e.g. L12, L13, L21) contained internal EcoR1 sites, and these cDNAs would have been fragmented had the library been constructed without the methylation step. A summary of the characteristics of one of the two sets of library constructions is given in Table I.

Table I: Features of a spinach cDNA library

Vector	Primary Library pfu/μg Poly(A^+)RNA	Amplified Library pfu/ml	volume
λgt10	4×10^6	1×10^{10}	10 ml
λgt11	8×10^6	1×10^{11}	15 ml

*Taken from the Doctoral Thesis of K. Giese (1990)

Protein Purification

The purification of several chloroplast ribosomal proteins have been described by us (Bartsch et al 1982, Kamp et al 1987, Johnson et al 1990; Smooker et al 1990a,b). Application of HPLC separation yielded purified individual proteins (e.g. L2, L21, L35, PSrp-1) from the pools of a previous large scale isolation procedure. The HPLC method has also been applied directly with the total proteins (TP30, TP50) extracted from chloroplast 30S and 50S subunits. Separation of chloroplast TP30 gave an elution profile with over 13 major peaks of which 7 peaks contained single proteins, 4 peaks contained 2 proteins each and 2 peaks contained 3 proteins each, covering a total of 21 proteins. In the case of TP50, over 21 peaks were obtained of which 13 contained single proteins, 6 contained 2 proteins each and 2 contained 3 proteins each, covering a total of 31 proteins. The separations were done on columns of Vydac RPSC C18 and C3 (A.R. Subramanian, unpublished results).

Amino acid sequence analysis has allowed the identification of the purified proteins, in many cases, as chloroplast homologues of corresponding E. coli proteins. Examples are L2 (Kamp et al 1987), L12 (Bartsch et al 1982), L13 (Srinivasa, unpublished), L27 (A.R. Subramanian, unpublished) and L35 (Smooker et al 1990b). In several other cases, however, the amino acid sequences showed no homology to any E. coli ribosomal protein. For some (e.g. L21) this arose because the protein sequence from the N-terminus belongs to an overhang that is absent in the corresponding E. coli protein. In two cases, however, there was extensive sequence data including sequences from internal peptides, but still no homology was found. The cDNA for one of these proteins (PSrp-1) has since been isolated and characterized. This chloroplast protein does not have a homologue in the E. coli ribosome, and we have proposed that it may represent an acquisition in the post-endosymbiont evolution of the chloroplast ribosome (Johnson et al 1990).

Immunological Relationships

Antisera have been raised to several chloroplast ribosomal proteins using either purified proteins or pools of partially purified material. These antisera have allowed the isolation of cDNA clones for proteins L11 (Smooker et al, Ms in preparation), L13 (Phua et al 1989), L21 (Smooker et al 1990a) and PSrp-1 (Johnson et al 1990) by screening the expression library in λgt11.

In a previous study from our laboratory antisera against 35 E. coli ribosomal proteins were tested for specific cross reactions to spinach chloroplast ribosomal proteins (Bartsch et al 1982; Bartsch 1985). Now that the complete list of the ribosomal proteins encoded in the chloroplast DNA of rice, tobacco and liverwort is known, results from the earlier study can be reviewed <u>vis à vis</u> nuclear versus plastid encoded proteins. Of the 35 proteins involved in this study, 11 are now known to be plastid-encoded and therefore the remaining 24 can be assumed to be nuclear (Table II). We detected a total of 12 crossreactions, 5 in the plastid category (S7, S11, S12, S19 and L2) and 7 in the nuclear category (S9, L1, L3, L6, L12, L13 and L17). To the latter must be added S1 to which cross reaction was detected by using monoclonal antibodies (Hahn et al 1988).

Table II: Chloroplast ribosomal proteins of higher plants: gene location and antibody crossreaction data

Ribosomal protein	Gene location	Crossreaction
S1	Nucleus	+
S2	Chloroplast	not tested
S3	Chloroplast	–
S4	Chloroplast	–
S5	Nucleus	–
S6	Nucleus	–
S7	Chloroplast	+ (43)[a]
S8	Chloroplast	–
S9	Nucleus	+
S10	Nucleus	–
S11	Chloroplast	+ (48)[a]
S12	Chloroplast	+ (66)[a]
S13	Nucleus	not tested
S14	Chloroplast	–
S15	Chloroplast	not tested
S16	Chloroplast	–
S17	Nucleus*	–
S18	Chloroplast	not tested
S19	Chloroplast	+ (42)[a]
S20	Nucleus	–
S21	Nucleus	not tested
L1	Nucleus	+
L2	Chloroplast	+ (49)[a]
L3	Nucleus	+
L4	Nucleus	–
L5	Nucleus	–
L6	Nucleus	+
(L7)	Modified form of L12	
(L8)	Complex of L10(L12)4	
L9	Nucleus	–
L10	Nucleus	–
L11	Nucleus*	–

Ribosomal protein	Gen location	Crossreaction
L12	Nucleus*	+ (50)[a]
L13	Nucleus*	+ (56)[a]
L14	Chloroplast	not tested
L15	Nucleus*	–
L16	Chloroplast	not tested
L17	Nucleus	+
L18	Nucleus	–
L19	Nucleus	–
L20	Chloroplast	not tested
L21	Nucleus*	not tested
L22	Chloroplast	–
L23	Chloroplast	not tested
L24	Nucleus	–
L25	Nucleus	–
(L26)	Same as S20	
L27	Nucleus*	not tested
L28	Nucleus	not tested
L29	Nucleus	–
L30	Nucleus	not tested
L31	Nucleus	not tested
L32	Chloroplast	not tested
L33	Chloroplast	not tested
L34	Nucleus	not tested
L35	Nucleus*	not tested
L36	Chloroplast	not tested
54		

*The nuclear gene location has been confirmed by isolating cDNA clones (see text). The genes for S16 and S21 in liverwort are, unlike in angiosperms, located in the nucleus and chloroplast, respectively. The L22 gene in pea is nuclear. Several of the nuclear located genes in higher plants (e.g. L3, L5, L6, etc.) are organelle encoded in algae (see other chapters in the book).

[a]Percent identity in amino acid sequence between the cross-reacting protein and its E. coli homologue (where data are available).

So far antibodies against 19 of the E. coli ribosomal proteins have not been tested for their crossreaction to chloroplast proteins (Table II). Probably some of them, those that are raised to proteins known to have high sequence homology (e.g. L14, L16, L36 with 52, 53 and 62% identical sequences), will give positive crossreactions. Amino acid sequence identities in the positive cases in Table II have values ranging from 43 to 66%. It appears from the limited data available that the nuclear-coded proteins are as similar to E. coli ribosomal proteins as are the chloroplast-encoded proteins. Many of the nuclear-coded proteins, however, unlike the chloroplast-encoded proteins, carry overhangs that have no similarity to any known sequences (see below).

cDNA Clones for Ribosomal Proteins

The cDNA clones we have so far characterized are those for 50S ribosomal subunit proteins L11, L12, L13, L21, L35 and for the 30S subunit protein designated Psrp-1 (i.e. plastid-specific ribosomal protein). In all cases the proteins are encoded in precursor forms carrying long transit peptide sequences at the N-termini. All these proteins except L11 have been isolated from the ribosome and their N-terminal sequences have been determined. Therefore the precise cleavage sites of their transit sequences are known. A summary of the data for these proteins and some basic features of the derived transit sequences are given in Tables 3 and 4.

Table III: Characteristics of six nuclear-coded ribosomal proteins identified at the cDNA level

Protein	Chain length*		Sequence identity to E. coli homologue	Peptide overhangs	
	Precursor	Mature		N-term.	C-term.
L11	224	~152	52%	~nil	12
L12	189	133	50%	4	0
L13	250	203	56%	52	9
L21	256	201	33%	66	30
L35	159	73	42%	2	6
PSrp-1	302	236	nil	–	–

References: L11, Smooker et al, Ms in preparation; L12, Giese and Subramanian 1989; L13, Phua et al 1989; L21 and L35, Smooker et al 1990a,b; PSrp-1, Johnson et al 1990.
*Given as the number of amino acid residues.

Table IV: Characteristics of deduced transit peptides

Protein	Length of transit	Unusual Composition Ser+Thr	Trp+Tyr	Cleavage site sequence
L11	~72	31%	nil	not determined
L12	56	34%	"	RPIA↓AVEA
L13	47	31%	"	AVSA↓KKST
L21	55	22%	"	LPVA↓AKRR
L35	86	38%	"	TVFA↓AKGY
PSrp-1	66	30%	"	VVCM↓SWDG

Chloroplast proteins L13 and L21 are striking for the overhangs, i.e. peptide extensions at the ends of the homologous *E. coli* sequences, which they carry (Table III). L13 has an N-terminal overhang of 52 amino acid residues while L21 carries large overhangs at both termini (66/33 residues). The larger size of chloroplast L13 was already known from antibody crossreaction experiments (Bartsch 1985). The C-terminal overhang of L21 has the sequence:

172 YEDYPASTLEAEVEAKEEAEAEAEAEAVPV 201

which contains 10 negatively charged residues (E = glutamic acid) and includes 7 repeats of the dipeptide Glu-Ala (EA). Since highly negatively charged regions are important in nucleic acid binding regions of several regulatory proteins this overhang will probably have an important functional role.

cDNA clones for the spinach proteins L21 and PSrp-1 have also been isolated by Mache and coworkers (T. Langrange et al, data presented at the meeting; Zhou and Mache 1989 and 1990). In tobacco, Bourque and coworkers have isolated cDNA clones of L12, L24 and L27, and in pea Gantt and coworkers have isolated cDNA clones for L15 and L22 (data presented at the meeting). Taking into account the earlier work from Gantt's lab (Gantt 1988, Gantt and Thompson 1990), the list of characterized cDNA clones include 12 proteins of the large ribosomal subunit and two proteins of the small subunit. Eleven of them are homologues of E. coli proteins, i.e. L9, L11, L12, L13, L15, L21, L22, L24, L27, L35 and S17, while PSrp-1 and two others (Gantt 1988) have no homologues on the bacterial ribosome.

Proteins L9, L11, L12, L13, L15, L21 and L22 are rRNA binding proteins which are present in the intermediate particle (RI_{50}) in the first step of E. coli 50S subunit assembly; L13 and L22 are among the 5 or 6 essential proteins of this step (Herold and Nierhaus 1987). Protein S17 is also an assembly protein, essential for RI_{30} formation. Thus many of the proteins essential for ribosome assembly are encoded in the nucleus. As we pointed out earlier in an Introduction (Prombona and Subramanian 1989), 16 of the ribosomal proteins encoded in the plastid DNA are assembly proteins. Thus, although the rRNA is encoded in the plastid DNA, the proteins required for the early steps of ribosome assembly are encoded in both the plastid and the nuclear genomes.

Organelle to Nucleus Gene Relocation

Two of the nuclear-coded proteins I have discussed are of special interest from an evolutionary point of view. Proteins L21 and L35 are nuclear-coded in spinach (our work) but organelle-encoded in certain lower species. One can therefore argue that the genes for these proteins must have been relocated into the nucleus more recently than at the initial phase of endosymbiosis.

As discussed in our primary publications (Smooker et al 1990a,b), the relocation of the rpl21 gene would have occurred in the intervening period between the appearances of Bryophyta and Angiospermae. This gene is nuclear in spinach (and presumably other flowering plants) but organelle-encoded in liverwort and presumably other members of the division Bryophyta. From fossil data (Stewart 1983), Bryophyta first appeared $350-400x10^6$ years ago. The angiosperms are believed to have appeared $150-200x10^6$ years ago (Stewart 1983; Doyle and Donoghue 1987). Thus the organelle to nucleus relocation event for rpl21 gene occurred later than $400x10^6$ but earlier than

150×10^6 years ago. The L35 gene is nuclear in spinach as well as liverwort but is organelle-encoded in the aquatic photosynthetic organism Cyanophora paradoxa. The nuclear relocation event in this case must therefore have occurred before the emergence of land plants around 400×10^6 years ago.

The deduced amino acid sequence of the L21 protein of liverwort, which is plastid-encoded, lacks the N- and C-terminal overhangs (including the unusual Ala-Glu repeat) found in the nuclear-coded L21 of spinach. The addition of the overhangs thus appears linked with the plastid- to nuclear gene migration. Whether it also means the acquisition of an additional functional role for the nuclear-coded L21 protein remains to be seen.

References

Bartsch M (1985) Correlation of chloroplast and bacterial ribosomal proteins by cross-reactions of antibodies specific to purified Escherichia coli ribosomal proteins. J Biol Chem 260:237-241

Bartsch M, Kimura M and Subramanian AR (1982) Purification, primary structure, and homology relationships of a chloroplast ribosomal protein. Proc Natl Acad Sci USA 79:6871-6875

Dorne AM, Eneas-Filho J, Heizmann P and Mache R (1984) Comparison of ribosomal proteins of chloroplast from spinach and of E. coli. Mol Gen Genet 193:129-134

Doyle JA and Donghue MJ (1987) The origins of angiosperms: a cladistic approach. In "The Origins of Angiosperms and Their Biological Consequences" (Friis EM, Chaloner WG and Crane PR, eds), pp. 17-49. Cambridge University Press

Gantt JS (1988) Nucleotide sequences of cDNAs encoding four complete nuclear-coded plastid ribosomal proteins. Curr Genet 14:519-528

Gantt JS and Thompson MD (1990) Plant cytosolic ribosomal protein S11 and chloroplast ribosomal protein CS17. J Biol Chem. 265:2763-2767

Giese K (1990) Ph.D. Dissertation. Free University of Berlin.

Giese K and Subramanian AR (1989) Chloroplast ribosomal protein L12 is encoded in the nucleus: Construction and identification of its cDNA clones and nucleotide sequence including the transit peptide. Biochemistry 28:3525-3529

Hahn V, Dorne AM, Mache R, Ebel JP and Stiegler P (1988) Identification of an Escherichia coli S1-like protein in the spinach chloroplast ribosome. Plant Mol Biol 10:459-464

Herold M and Nierhaus KH (1987) Incorporation of six additional proteins to complete the assembly map of the 50S subunit from Escherichia coli ribosomes. J Biol Chem. 262, 8826-8833

Johnson CH, Kruft V and Subramanian AR (1990) Identification of a plastid-specific ribosomal protein in the 30S subunit of chloroplast ribosomes and isolation of the cDNA clone encoding its cytoplasmic precursor. J Biol Chem 265:12790-

12795

Kamp RM, Srinivasa BR, von Knoblauch K and Subramanian AR (1987) Occurrence of a methylated protein in chloroplast ribosomes. Biochemistry 26:5866-5870

Lyttleton JW (1962) Isolation of ribosomes from spinach chloroplasts. Exp Cell Res 26:312-317

Mache R, Dorne AM and Batlle RM (1980) Characterization of spinach plastid ribosomal proteins by two-dimensional gel electrophoresis. Mol Gen Genet 177:333-338

Phua SH, Srinivasa BR and Subramanian AR (1989) Chloroplast ribosomal protein L13 is encoded in the nucleus and is considerably larger than its bacterial homologue. J Biol Chem. 264:1968-1971

Prombona A and Subramanian AR (1989) A new rearrangement of angiosperm chloroplast DNA in rye (Secale cereale) involving translocation and duplication of the ribosomal rps15 gene. J Biol Chem 264:19060-19065

Smooker PM, Kruft V and Subramanian AR (1990a) A ribosomal protein is encoded in the chloroplast DNA in a lower plant but in the nucleus in angiosperms. Isolation of the spinach L21 protein and cDNA clone with transit and an unusual repeat sequence. J Biol Chem. 265:16699-16703

Smooker PM, Choli T and Subramanian AR (1990b) Ribosomal protein L35: Identification in spinach chloroplasts and isolation of a cDNA clone encoding its cytoplasmic precursor. Biochemistry, in press

Stewart WN (1983) Paleobotany and the evolution of plants. Cambridge University Press

Subramanian AR, Smooker PM and Giese K (1990) Chloroplast ribosomal proteins and their genes. In "The Ribosome: Structure, Function, & Evolution (Hill WE, Dahlberg A, Garrett RA, Moore PB, Schlessinger D and Warner JR, eds). American Society for Microbiology Publications, Washington DC

Subramanian AR, Stahl D and Prombona A (1991) Ribosomal proteins, ribosomes and translation in plastids. In "The Molecular Biology of Plastids and the Photosynthetic Apparatus" (Bogorad L and Vasil IK, eds) Academic Press, in press

Zhou DX and Mache R (1989) Presence in the stroma of chloroplasts of a large pool of a ribosomal protein not structurally related to any Escherichia coli ribosomal protein. Mol Gen Genet 219:204-208

Zhou DX and Mache R (1990) Erratum: Presence in the stroma of chloroplasts of a large pool of a ribosomal protein not structurally related to any Escherichia coli ribosomal protein. Mol Gen Genet 223:167

Comparative analysis of four different cDNA clones encoding chloroplast ribosomal proteins

T. Lagrange, P. Carol, C. Bisanz-Seyer and R. Mache
Laboratoire de Biologie Moléculaire végétale, Université Joseph Fourier and CNRS, BP 53X, F-38041 Grenoble Cedex, France

Introduction

Chloroplast ribosomes consist of about 60 different rRNA and protein components and their biogenesis depends on the expression of genes located within both the chloroplast and nuclear genomes. The endosymbiotic theory predicts that all chloroplast ribosomal proteins were initially encoded within the chloroplast DNA. During the course of evolution, about two thirds of the chloroplast ribosomal proteins (r-proteins) genes have been transferred to the nucleus (Dorne et al 1984; Sugiura, 1989). In order to better understand the processes which have driven the compartmentalisation of chloroplast ribosomal protein genes we have isolated cDNA clones encoding chloroplast ribosomal proteins and have investigated their structure.

We give here a short comparative analysis of 4 cDNA clones from spinach, each of which encodes a different chloroplast r-protein. It is shown that these nuclear encoded ribosomal proteins contain new structural elements not found in *E. coli* ribosomal proteins, thus providing evidence that prokaryotic ribosomes have been deeply remodelled during the course of evolution of plant cells, mainly through the sequestering of r-protein genes in the nucleus.

Nomenclature

The standard nomenclature system concerning chloroplast ribosomal proteins is based on sequence similarity with *E. coli* r-proteins (Hallick and Bottomley, 1983). Thus, two of the r-protein genes reported here, *rpl* 21 and *rpl* 24, are named according to the sequence similarities of the respective coding regions to the *E. coli* L21 and L24 r-proteins. Two other r-proteins

NATO ASI Series, Vol. H 55
The Translational Apparatus of Photosynthetic Organelles
Edited by R. Mache et al.

described here have no *E. coli* counterparts and are called chloroplast specific r-proteins (Zhou and Mache, 1989). For these proteins we propose use of the same nomenclature system, but with extension of serial numbers used in *E. coli*. Twenty one r-proteins have been identified in the *E. coli* small ribosomal subunit and consequently the first chloroplast r-protein not homologous to an *E. coli* r-protein is named S22 (see Table 1). In the case of the *E. coli* large ribosomal subunit, 36 r-proteins have been described so far. But as the number of r-proteins is not known with certainty, the first non homologous protein is named L40.

Table 1. Nomenclature of the chloroplast ribosomal proteins described in this paper (see text)

gene	gene product	other name	References
rpl 21	L21	CS-L7	Mache et al. 1980; Martin et al.1990; Smooker et al. 1990
rpl 24	L24	CS-L19	Mache et al 1980; Carol et al. submitted
rpl 40	L40	Ps-L18	Gantt 1988; Carol et al. submitted
rps 22	S22	CS-S5	Zhou and Mache 1989; Zhou and Mache 1990
		PSrp-1	Johnson et al. 1990

Comparative analysis of cDNAs encoding the spinach chloroplast r-proteins L21, L24, L40 and S22

Four different cDNA clones, each encoding a chloroplast 70S r-protein have been isolated from a spinach λgt11 library. The nucleotide sequences of these cDNAs are reported elsewhere (Zhou and Mache 1989, 1990; Martin et al. 1990; Carol et al. submitted) and are available in the EMBL data bank.The main features of the four cDNAs as well as the characteristics of the encoded proteins will be discussed here. Two different cDNAs code for proteins homologous to the L21 and L24 *E.coli* ribosomal proteins, respectively, while two other cDNAs code for chloroplast r-proteins (L40 and S22; see also Table 1) which have no significant homology with any of the *E.coli* r-proteins.

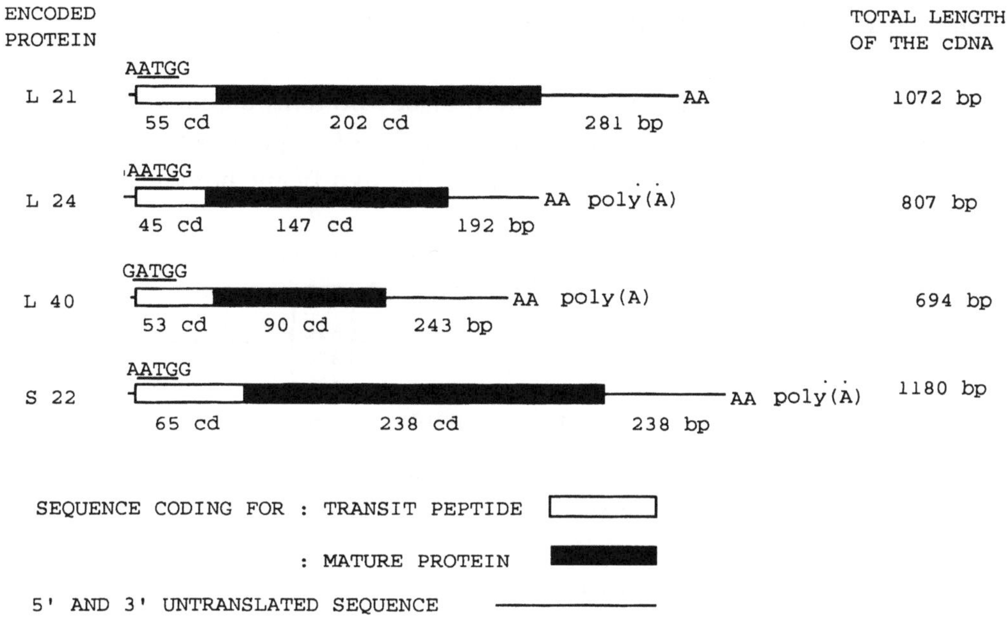

Fig 1: Schematic representation of the cDNAs encoding spinach chloroplast ribosomal proteins L21, L24, L40 and S22. White bars represent the transit peptides and black bars the mature proteins.The indicated number of codons (cd) includes the stop codon. The length of the 3' non translated region is given in base pairs (bp).

Figure 1 gives a schematic comparison of the cDNAs encoding r-proteins L21, L24, L40 and S22, respectively. A Kozak consensus sequence, PuXXATGG, is found at the starting translational site of the four cDNAs. Such a sequence is generally found at the initiation site for translation of plant cytosolic mRNAs (Lütcke et al 1987). A poly(A) tail is observed at the 3' end of the four cDNAs but no consensus polyadenylation sites (AAUAA) are present. By sequencing the 3' non coding regions of two cDNA clones encoding the S22 r-protein, we found that the poly(A) sequences were located at different positions in the sequence. The presence of multiple polyadenylation sites has been previously observed (Hernandez-Lucas et al 1986; Dean et al 1985) and evidence was obtained for mRNAs in animal cells suggesting that the position of polyadenylation sites have a regulatory function in the control of gene expression (Birnstiel et al. 1985). The significance and function of multiple polyadenylation sites for *rps*22 transcripts is at present unknown.

Composite structure of ribosomal proteins L21 and L24

Many of the nuclear encoded chloroplast r-proteins curently known are composed of a central core homologous to an *E. coli* r-protein surrounded by N- and C-terminal extensions which have no significant homology to any bacterial r-protein (Mache 1990).

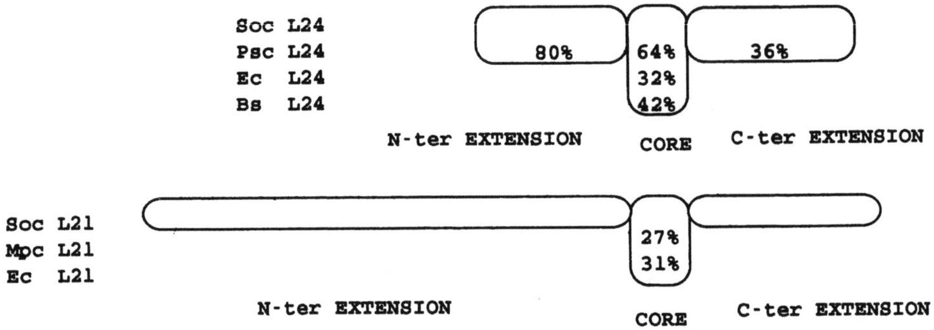

Fig. 2: Amino acid alignments of spinach L21 (**A**) and L24 (**B**) ribosomal proteins with homologous r-proteins from other species: Soc, spinach; Mp, *Marchantia polymorpha* (Kochi et al 1988); Ec, *Escherichia coli* (Heiland and Wittmann-Liebold 1979 (L21) and Wittmann-Liebold 1979 (L24); Psc, pea (Gantt 1988) and Bs, *Bacillus stearothermophilus* (Kimura et al 1980).The N- and C-terminal extensions of the mature higher plant r-proteins are given in detail. Percentages of homology relative to spinach are indicated within a block representing the central core of the proteins.

Figure 2 shows the amino acid alignments of the mature spinach L21 (A) and L24 (B) r-proteins with their homologues from other species. The N-terminal extensions of the spinach and pea L24 r-proteins share a high percentage of amino acid identity (80%) wheras the C-terminal extensions are homologous by 37% only. These results show that extensions were present in the nuclear gene before the divergence of legumes from other higher plants. We propose that the coding sequences for polypeptide extensions have been added to r-protein genes at the time of transfer of these genes from the chloroplast genome to the nucleus. The length of extensions is variable, with up to 67 amino acids in the case of the N-terminal extension of the spinach L21 protein

(Martin et al,1990). The function and the origin of these extensions are not known but their existence give an evidence that remodelling of prokaryotic ribosomes has occurred during the course of evolution of land plants.

Chloroplast specific ribosomal proteins

No sequence similarities between the nuclear-encoded chloroplast r-proteins L40 and S22 and any *E.coli* r-protein has been found. Thus it seems that some r-proteins have been lost from the primordial cyanobacterial ribosome during the course of evolution and that they have been replaced by ribosomal proteins of yet unknown origin. It would be very interesting to search for the presence of r-proteins similar to L40 or S22 in archaebacteria or in eucaryotes.

Comparison of transit peptides

Nuclear-encoded plastid proteins are synthesized as precursors in the cytoplasm and are imported post-translationally into the organelle (Keegstra 1989). The N-terminal amino acid sequencing of the mature L21, L40 and S22 ribosomal proteins (Martin et al, 1990; Carol et al, submitted; Johnson et al, 1990) has allowed the identification of the cleavage site of the transit peptide of each of these proteins. Recently, Gavel and von Heijne (1990) proposed a cleavage site consensus sequence for higher plant transit peptides: (Val or Ile)-X-(Ala or Cys) Ala, with in addition, arginines in positions -6 to -10. These characters are verified for the three ribosomal proteins L21, L40 and S22. The cleavage site of the L24 is not yet determined. It could be located within the CLIVM amino acid sequence which is common to the pea and spinach L24 protein because this region is preceeded by 2 or 3 charged amino acid residues as generally observed (von Heijne et al.1989).

Proteins are stoichiometrically represented in ribosomes and consequently the same amount of each chloroplast r-protein synthesized in the cytoplasm has to cross the chloroplast envelope. One can also imagine that r-proteins which are not incorporated into ribosomes are degraded. In such a case, an excess of some r-proteins could be transported before degradation. We have examined some common features in the transit peptides of the nuclear encoded ribosomal proteins. The amino sequence comparison which is shown in Fig. 3 show that the four transit peptides possess the same properties as other nuclear encoded chloroplast proteins (von Heijne et al 1989; Keegstra and Olsen 1989). For instance, they have an uncharged amino-terminal domain, the residue next to the initiator Met is Ala, they have a high content of Ser and Thr residues and the carboxy-terminal region contain an increased porportion of positively charged residues. Also observed is a deficiency in acidic amino acids. No common amino acid sequences are found in all four transit peptides.

```
 o   oo          o o o o    ooo o   oo +oo o o o          +o    o+ +
MATLCTSAINMNPNLTNSLSNSINLSSTLTNLSSLRSTFTNSCSLGLNVAVKSVQISRNKPNVVCM S   TP S22

                   o o   o oo   o +                o+  o    o   o o+ o
                  MASATLAFSCSSLCATLKLPQNLNPLLLNVPPLSKPFSGVVSPPSLSRLSLLPVA A   TP L21

                               oo oo o oo o     +   o oo    +o- +o
                            MAAMVLQSSFTSLSLSSNSFLGQRLFPSPTTLQVKTEGHSPCLIV M   TP L24

      o   o oo    oo   oooo   +    o      o    +    +       ++ o
    MALLSPLLSLSSVPPITSIAVSSSSFPIKLQNVSVALLPSFGQRLVAHGPVIAQKRGTVVAM V   TP L40
```

Fig. 3: Transit peptide (TP) sequences of the chloroplast S22, L21, L24 and L40 ribosomal proteins. Aminoacids with hydroxyl (o), positively charged (+) and negatively charged (-) chains are indicated. The conserved sequence MA is underlined. The processing site of the precursor L24 is putative.

Nuclear genes for chloroplast ribosomal proteins L21, L40 and S22 are present at a low copy number

Southern genomic experiments were made to determine the number of copy of the nuclear genes coding for the chloroplast L21, L40 and S22 ribosomal proteins, respectively. Results have shown that in all cases only one gene is present in the nuclear genome (Martin et al, 1990 and unpublished results). This result is in contrast with the presence of 80S r-protein gene families containing 10 to 25 members in mamalian genomes (Wagner and Perry 1985).

Conclusions

The comparative analysis of four nuclear encoded chloroplast ribosomal proteins in spinach has shown that, despite many similarities between chloroplastic and prokaryotic ribosomes, there are important differences. In particular, the presence of chloroplast specific ribosomal proteins and of ribosomal proteins composed of a central homologous core and of extensions show that large modifications have occurred in the architecture of chloroplast ribosomes. The reason of these modifications might be an adaptation of the translational apparatus to environmental conditions which are very different from those of the primitive cyanobacterial endosymbiont. It is worth noting that the chloroplast encoded L21 r-protein in the liverwort *Marchantia polymorpha* has no extensions as those described here for spinach, suggesting that modifications of chloroplast ribosomes have occurred during the evolution of higher plants. The comparison of the L24 r-protein in pea and spinach indicate that the "polypeptide extensions" were added before the divergence of legumes from other higher plants, i.e. more than 70-80 millions years ago. We may extend these observations by suggesting that remodelling of chloroplast ribosomes made by the introduction of specific amino acid sequences (extensions or entire proteins) has also occured in land plants more than 70-80 million years ago.

Acknowledgement. We thank Dr WF Martin for critical reading of the manuscript.

References

Birnstiel ML, Busslinger M, Strub K (1985) Transcription termination and 3' processing: the end is in site. Cell 41, 349-359

Dean C, Van den Elzen P, Tamaki S, Dunsmuir P, Bedbrook J (1985) Differential expression of the eight genes of the petunia ribulose bisphosphate carboxylase small subunit multi-gene family. EMBO J 4, 3055-3061

Dorne AM, Eneas-Filho J, Heizmann P, Mache R (1984) Comparison of ribosomal proteins of chloroplasts from spinach and of *E. coli* . Mol Gen Genet 193, 129-134

Gantt JS (1988) Nucleotide sequences of cDNAs encoding four complete nuclear-encoded plastid ribosomal proteins. Curr Gen 14, 519-528

Gavel Y, Von Heijne (1990) A conserved cleavage-site motif in chloroplast transit peptides. FEBS Lett 216, 455-458

Hallick RB, Bottomley W (1983) Proposals for the naming of chloroplast genes. Plant Mol Biol Rep 1, 38-43

Heiland I, Wittmann-Liebold B (1979) Amino acid sequence of the ribosomal protein L21 of *Escherichia coli.* Biochem 18, 4605-4612

Hernandez-Lucas C, Royo J, Paz-Ares J, Ponz F, Garcia-Olmedo F, Carbonero P (1986) Polyadenylation site heterogeneity in mRNA encoding the precursors of the barley toxin *b*-hordothionin. FEBS Lett 200, 103-106

Johnson CH, Kruft V, Subramanian AR (1990) Identification of a plastid-specific ribosomal protein in the 30S subunit of chloroplast ribosomes and isolation of the cDNA clone encoding its cytoplasmic precursor. J Biol Chem 265, 12790-12795

Keegstra K, Olsen LJ (1989) Chloroplastic precursors and their transport across the envelope membranes. In: Ann Rev Plant Physiol Plant Mol Biol 40, 471-501

Kimura M, Dijk J, Heiland I (1980) The primary structure of protein BL17 (= homolog of Eco-L9) isolated from the large subunit of the *Bacillus stearothermophilus* ribosome. FEBS Lett 121, 323-326

Kohchi T, Shirai H, Fukuzawa H, Sano T, Komano T, Umesono K, Inokuchi H, Ozeki H, Ohyama K (1988) Structure and organization of Marchantia polymorpha chloroplast genome. IV. Inverted repeat and single copy regions. J Mol Biol 203, 353-372

Lütcke HA, Chow KC, Mickel FS, Moss KA, Kern HF, Scheele GA (1987) Selection of AUG initiation codons differs in plants and animals. EMBO J 6, 43-48

Mache R, Dorne AM,Marti-Battle (1980) Characterization of spinach plastid ribosomal proteins by two dimensional gel electrophoresis. Mol Gen Genet 177, 333-338

Mache R (1990) Chloroplast Chloroplast ribosomal proteins: a review. Plant Science (in press)

Martin W, Lagrange T, Li YF, Bisanz-Seyer C, Mache R (1990) Hypothesis for the evolutionary origin of the chloroplast ribosomal protein L21 of spinach. Curr Genet (in press)

Smooker P, Kruft V, Subramanian AR (1990) A ribosomal protein is encoded in the chloroplast DNA in a lower plant but in the nucleus in angiosperms. J Biol Chem (in press)

Sugiura M (1989) The chloroplast chrosomes in land plants. Ann Rev Cell Biol 5: 51-70

Von Heijne G, Stepphuhn J, Herrmann RG (1989) Domain structure of mitochondrial and chloroplast targeting peptides. Eur J Biochem 180: 535-545

Wagner M, Perry RP (1985) Characterization of the multigene family encoding the mouse S16 ribosomal protein: strategy for distinguishing an expressed gene from its processed pseudogene counterparts by analysis of total genomic DNA. Mol Cell Biol 5, 3560-3576

Wittmann-Liebold B (1979) Primary structure of protein L24 from the *Escherichia coli* ribosome. FEBS Lett 108, 75-80

Zhou DX, Mache R (1989) Presence in the stroma of chloroplasts of a large pool of a ribosomal protein not structurally related to any *Escherichia coli* ribosomal protein. Mol Gen Genet 219, 204-208

Zhou DX, Mache R (1990) Presence in the stroma of chloroplasts of a large pool of a ribosomal protein not structurally related to any *Escherichia coli* ribosomal protein. *Erratum*. Mol Gen Genet 223, 167

TWO GENE FAMILIES ENCODING CYTOPLASMIC RIBOSOMAL PROTEINS IN BARLEY

K. Gausing, C. B. Jensen, J. D. Kreiberg, and L. H. Madsen
Department of Molecular Biology and Plant Physiology
University of Aarhus
DK-8000 Århus C
Denmark

INTRODUCTION

Biogenesis of cytoplasmic ribosomes in higher plants was initially followed *via* the accumulation of rRNA. Maximal levels of rRNA was seen in developing endosperm at the end of the period of rapid cell division (Müntz, 1982). Ribosomal RNA synthesis is induced very early during germination (Bewley, 1982), and rRNA accumulation in the first leaves of wheat and barley is most vigorous in the basal meristem and young leaf cells (Dean and Leech, 1982, Barkardottir et al., 1987). The link between cell proliferation and ribosome biogenesis was also observed in soybean hypocotyl hooks induced with auxin where a several fold increase of r-protein mRNAs was observed (Gantt and Key, 1985). Recently cDNA clones encoding plant r-proteins have been isolated (Larkin et al., 1989, Gantt and Thompson, 1990), the structure of rDNA has been studied, and rRNA transcripts have been mapped (Vincentz and Flavell, 1989) opening the way to a much better understanding of control of r-gene expression as well as ribosome biogenesis. We are interested in the developmental control of expression of housekeeping genes in germinating embryos and young seedlings and here summarize our results on the structure and expression of two genes encoding ubiquitin-long tail fusion peptides (Gausing and Jensen, 1990) and cDNA clones representing two genes encoding r-proteins homologous to yeast r-protein L17 (Madsen et al., submitted).

NATO ASI Series, Vol. H 55
The Translational Apparatus of Photosynthetic Organelles
Edited by R. Mache et al.

THE *mub* GENE FAMILY

There are two kinds of ubiquitin genes in eukaryotes: polyubiquitin genes encoding tandem repeats of the ubiquitin monomer and monoubiquitin genes encoding ubiquitin with short or long basic extensions called tails. Recently is was discovered that the tails are r-proteins (Finley et al., 1989, Redman and Rechsteiner, 1989,) and results were presented suggesting that the role of ubiquitin is to act as a chaperone directing the r-proteins to the ribosomal particles (Finley et al., 1989).

Using a ubiquitin cDNA as probe (Gausing and Barkardottir, 1986) a barley genomic DNA fragment that contains two monoubiquitin-long tail genes, *mub1* and *mub2*, has been cloned (Fig. 1). Both genes are active and give rise to messengers about 800 nucleotides long. The two genes are very homologous (96% in the coding regions) and they probably arose by a recent gene duplication. A probe specific for the long tail coding region of the *mub* genes was used for a Southern hybridization of genomic DNA and the result showed that there most likely are three long tail genes in the barley haploid genome (Gausing and Jensen, 1990).

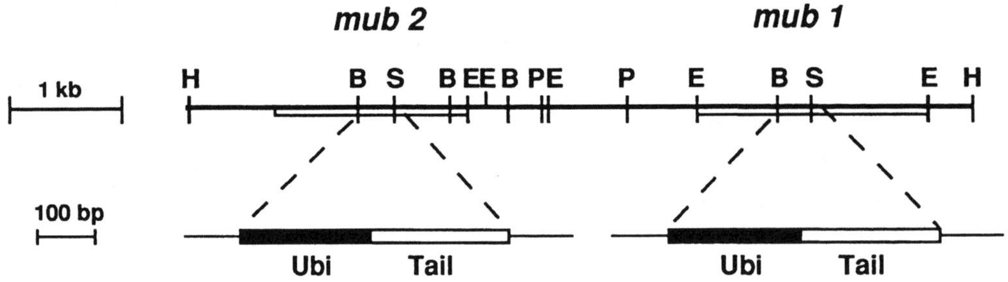

Fig. 1. Map of barley genomic DNA fragment with two genes encoding monoubiquitin-long tail fusions. The top line shows a 7.2 kb HindIII fragment with the regions that have been sequenced indicated with open boxes. The GenBank accession numbers assigned to the sequences are M34707-mub1 and M34708-mub2. The second line shows the extent of the primary transcripts as determined by S1 nuclease mapping. The open reading frames for ubiquitin and the tail peptides are indicated with black and open boxes, respectively. The abbreviated restriction enzymes sites are: *Bgl*II (B), *Eco*RI (E), *Hind*III (H), *Pst*I (P), and *Sal*I (S).The barley genomic library (cv Bomi) was a kind gift from A. Brandt, Carlsberg Laboratory. It was constructed by inserting partially *Hind*III digested DNA into λL47.1 DNA. The library was screened with insert from the polyubiquitin cDNA clone, pKG3730 (Gausing and Barkardottir, 1986) and among the recombinants purified was λUbi515 containing the 7.2 kb *Hind*III fragment shown.

The 5´untranslated leader sequences of the *mub* mRNAs have an exceptionally high CG content, 80-81%. Most leader sequences both in plant and animal messengers are AU rich and the CG content is rarely above 50% (Joshi, 1987). The partial leader sequences of barley L17 mRNAs (see below) as well as the partial leader sequences published for maize S14 mRNA (Larkin et al., 1989) and Arabidopsis S11 mRNA (Gantt and Thomson,1990), are also CG rich. However, it is probably too early to decide if CG rich leaders are characteristic of plant ribosomal protein genes in general.

The promoter/regulatory regions of the *mub* genes were compared with upstream regions of animal (mouse, Hariharan et al., 1989) and yeast ribosomal protein genes (review: Planta and Raué, 1988). The upstream regions from both organisms have been mapped with deletion analyses and factor binding experiments. There is no similarity between the animal and yeast upstream regions but there is similarity between the yeast and plant upstream regions. The promoter regions of the *mub* genes contain sequences with a 12/15 match with yeast UASrpg (upstream activating sequences for ribosomal protein genes, Planta and Raué, 1988) , and the location of these sequences is comparable in barley and yeast.

The sequence of the ubiquitin moiety encoded by the *mub* genes is identical to the sequence of ubiquitin repeats of polyubiquitin precursors from other plants. The basic tails are 79 amino acids long and 71-72% homologous to corresponding sequences from yeast and man (Fig. 2). The high cross-kingdom sequence homology between the long tail peptides as well as their fusion to ubiquitin suggest that the tails are ribosomal proteins in all eukaryotes (Finley et al., 1989).

```
                                                        40
Barley mub1    AKKRKKKTYTKPKKQKHKHKKVKLAVLQFYKVDDATGKVT
Barley mub2    ----------------------------------S-----
Yeast          G------V--T---I-------------SY----A*E----
Human          -------S--T---N---R---------KY----E*N--IS

                                                       79
Barley mub1    RLRKECPNADCGAGTFMANHFDRHYCGKCGLTYVYNQKA
Barley mub2    ---------------------------------------
Yeast          K--R--S-PT----V-L---K--L-----HSV-KV-A
Human          ---R---S-------VR--S-----------C---CF-KPEDK
```

Fig.2. Sequences of the tails of ubiquitin-long tail fusion peptides in barley, yeast (Özkaynak et al., 1987), and man (Lund et al., 1985). Identical amino acids are shown by dashes and deletions by stars.

```
                            1                 10                  20                  30                  40
H. marismortui L23          M G I S Y S V E - A D P D T T A K A M L R E R Q M S F K H S K A I A R E I K G K T A G
YL17                            A R Y G A T S T N P A K S A S A R G S Y L R V S F K B T R Z T A Z A I B G X Z L
HvL17-1 (pKG1973)             M V K Y S T D P A N P T K S A K A M G R D L R V H F K N T R E T A F A L R K M P L N
Liverwort cp L22                        M Q T N T S N K K I R A V A K H I H M S P H K V R R V V S Q I R G R S Y E
E. coli L22                                           M E T I A K H R H A R S S A Q K V R L V A D L I R G K K V S

                     50                  60                  70                  80                  90
H.m.  E A V D Y L E A V I E G D Q P V P F K Q H N S G V G H K S K V D G W D A - - - G R Y P E K A S K A F L D L
Hv1   K A K R Y L E D V L A H K Q A I P F R R Y C R G V G R T A Q V K N R Q P N G Q G R W P A K S A K F V L D L
Lw    Q A L M I L E F M - - - - - - - - - - - - - - - - - - - - - - - - - - - - - - - - P Y R A C N P I L Q L
E.c.  Q A L D I L T Y T - - - - - - - - - - - - - - - - - - - - - - - - - - - - - - - - N K K A A V L V K K V

                100                 110                 120                 130                 140
H.m.  L E N A V G N A D H - Q G F D G E A M T I K H V A A H K V G E Q Q G R K P R A M G R A S A W N S P Q V D V
Hv1   L K N A E S N A E V - K G L D V D A L Y I S H I Q V N Q A Q K Q R R R T Y R A H G R I N P Y M S N P C H I
Lw    L S S A A A N A N H N F G L S K T N L F I S E I Q V N K G T F F K R F Q P R A Q G R G Y P I H K P T C H I
E.c.  L E S A I A N A E H N D G A D I D D L K V T K I F V D E G P S M K R I M P R A K G R A D R I L K R T S H I

      150                 160                 170
H.m.  E L I L E E P E V E D
Hv1   E L I L S E K E E P V K K E A E S Q I A R K A
Lw      T I V L N I L P K
E.c.  T V V V S D R
```

Fig. 3. Alignment of one of the barley L17 amino acid sequences (HvL17-1) with the yeast L17 N-terminal fragment (Otaka et al.,1984), one of two known archaebacterial L22, one of several known chloroplast L22, and one of two known bacterial L22 proteins. The references are listed in the legend to Table 1. Identical amino acids that include the barley L17-1 sequence are boxed. The standard single letter code is used (B is D or N, Z is E or Q, and X is unknown).

TWO DIFFERENT TYPES OF cDNA SEQUENCES ENCODING R-PROTEINS HOMOLOGOUS TO YEAST R-PROTEIN L17

In a survey of our reference barley leaf cDNA library, a group of four cDNA clones representing genes preferentially expressed in meristematic and young cells was characterized (Barkardottir et al. 1987). The clones have inserts with two different homologous sequences, represented by pKG1973 and pKG989. Neither of the cDNA clones were full length but such clones were obtained from a second, larger library. 19 clones were analysed: 12 were derived from the same gene as pKG1973 and 7 were derived from the same gene as pKG989. The DNA sequences of the full length clones have single, long open reading frames of 170 and 172 codons, respectively. A search of the PIR protein sequence database revealed that the barley genes encode proteins with homology to yeast r-protein L17. Only the N-terminal 40 residues of YL17 have been determined (Otaka et al., 1984) but of these 40 residues at least 19 are identical in the barley sequences. In addition the search showed that the barley sequences are homologous to archaebacterial r-protein L23 from *Halobacterium marismortui* (Hatakeyama et al., 1988) and L22 from *Halobacterium halobium* (Mankin, 1989) as well as liverwort

Table 1 Summary of L17/L22 homologies

%	Archaebacteria	Eukaryote	Chloroplast	Eubacteria
Archaebacteria	67	22-33	13-17	17-19
Eukaryote	23-34	48	12-20	16-18
Chloroplast	21-23	18-30	38-55	23-37
Eubacteria	24-28	26-29	31-40	53

Numbers above the diagonal result from comparisons including all gaps and unmatched N- or C- terminal residues. Numbers in and below the diagonal results from comparisons excluding all gaps. The sequence of yeast L17 is only included in the eukaryote/eukaryote comparison. The L22 sequences used in the calculation were from *Halobacterium marismortui* L23 (Hatakeyama et al., 1988) and *H. halobium* L22 (Mankin, 1989), from liverwort (Ohyama et al., 1986), *Euglena* (Christopher et al., 1988), tobacco (Tanaka et al., 1986), and rice chloroplast L22 (Hiratsuku et al., 1989) and from *E. coli* (Wittmann-Liebold and Greuer, 1980) and *Mycoplasma* L22 (Ohkubo et al., 1987).

chloroplast L22 (Ohyama et al., 1986) that all are homologous to eubacterial L22 sequences. Representative alignments are shown in Fig. 3 and the amino acid sequence homologies are summarized in Table 1. Chloroplast L22 is coded by the chloroplast genome except in legumes (Gantt, ibid) and in spinach the chloroplast *rpl22* gene was directly shown to produce a protein homologous to E. coli L22 (Zhou et al. 1989).

As seen from the diagonal numbers in Table 1 the L17/L22 proteins are not highly conserved. Since the full length of yeast L17 is not known the barley sequences provide the link between the eukaryotic, archaebacterial and chloroplast/eubacterial sequences. The archaebacterial sequences are clearly more related to the eukaryotic sequences than to the eubacterial sequences since they share the large insertion (Fig. 3). Archebacterial r-protein sequences are in general more related to eukaryotic than to eubacterial r-protein sequences (e.g. Köpke and Wittmann-Liebold, 1989). On the other hand the archaebacterial genes are nested in S10 like operons similar to the arrangement in the *E. coli* genome and the archaebacterial rRNAs are eubacterial-like. The phylogenetic position of the archaebacteria is therefore still uncertain.

We have also determined the sequence of the barley chloroplast *rpl22* gene (unpublished). It is 93% homologous to the rice gene (Hiratsuku et al., 1989); curiously, the cereal chloroplast L22 proteins show the least homology of all prokaryotic type L22 sequences to the barley L17 sequences.

EXPRESSION OF THE RIBOSOMAL PROTEIN GENES

We have previously found that the ubiquitin gene family gives rise to mRNA species between 0.7 and 2 kb in length. The level of the larger mRNAs vary approximately two-fold in different vegetative tissues and is independent of the cell age in barley leaves. In contrast, the 0.7 kb mRNAs, now known to encode ubiquitin-r-protein fusions, were predominantly expressed in dividing and young developing cells (Gausing and Barkardottir, 1986). The level of the barley L17 mRNAs was also determined in the leaf gradient of young barley leaves, in root tips, and in roots without the meristematic tips (Madsen et al., submitted). Like the ubiquitin-r-protein mRNAs, the L17 mRNAs are predominantly expressed in the basal meristem and very young leaf cells and the level decreases rapidly in older leaf cells. The L17 mRNA level is also high in root meristem but undetectable in mature root cells. An illustration of these expression patterns is shown in Fig. 4. In young barley leaves, the majority of the cytoplasmic rRNA accumulates in the meristematic zone and a further increase of 50% (average per nucleus) is seen in cells between 6 and 40 h old (Barkardottir et al., 1987). Thus the level of the ubiquitin-r-protein mRNAs and the L17 mRNAs is to a first approximation coordinated with the accumulation of cytoplasmic ribosomes.

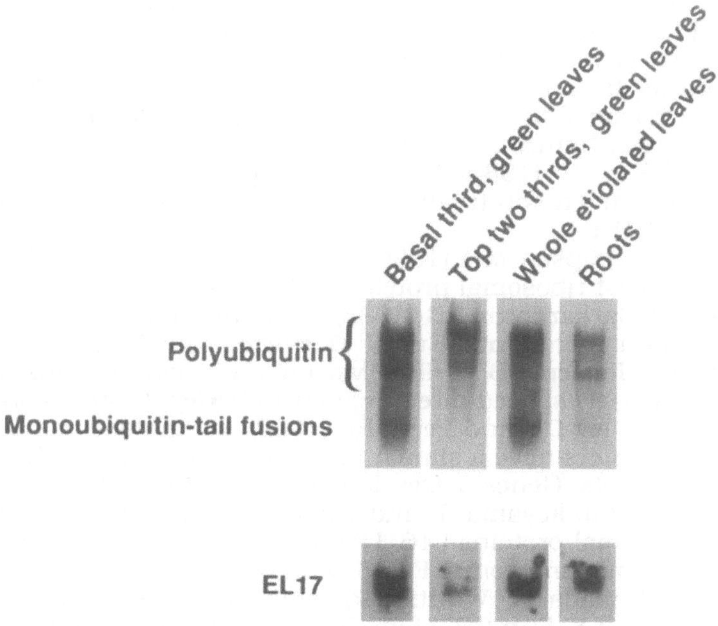

Fig. 4. Expression of ubiquitin genes and L17 genes in different barley tissues. Northern filters (Barkardottir et al. 1987) with total RNA samples (10 µg/lane) were hybridized with a ubiquitin insert from pKG3730 and a L17 insert from pKG1973. The basal one-third of 7 day old barley leaves includes cells up to the age of 36 h and most of the cells in the top two-thirds are between 1.5 and 5 days old. The whole etiolated leaves and the roots both contain meristems.

ACKNOWLEDGEMENTS

We wish to thank I. Bjørndal and L. Kjeldbjerg for expert technical assistance. The work was supported by The Danish Biotechnology Programme and a Carlsberg Foundation fellowship (LHM).

REFERENCES

Barkardottir RB, Jensen BF, Kreiberg JD, Nielsen, PS and Gausing K (1987) Expression of selected nuclear genes during leaf development in barley. Developmental Genetics 8: 495-511

Bewley JD (1982) Protein and nucleic acid synthesis during seed germination and early seedling growth. In Pirson A and Zimmermann MH (eds) Encyclopedia of Plant Physiology. New Series, Vol 14A Springer-Verlag Berlin Heidelberg New York pp 559-591

Christopher DA, Cushman JC, Price CA and Hallick RB (1988) Organization of ribosomal protein genes *rpl23*, *rpl2*, *rps19*, *rpl22*, and *rps3* on the *Euglena gracilis* chloroplast genome. Curr Genet 14: 275-286

Dean C and Leech R (1982) Genome expression during normal leaf development. Plant Physiol. 69: 904-910

Finley D, Bartel B and Varshavsky A (1989) The tails of ubiquitin precursors are ribosomal proteins whose fusion to ubiquitin facilitates ribosome biogenesis. Nature 338: 394-401

Gantt JS and Key JL (1985) Coordinate expression of ribosomal protein mRNAs following auxin treatment of soybean hypocotyls. J. Biol. Chem. 260: 6175-6181

Gantt JS and Thompson MD (1990) Plant cytosolic ribosomal protein S11 and chloroplast ribosomal protein CS17. J. Biol. Chem. 265: 2763-2767

Gausing K and Barkardottir RB (1986) Structure and expression of ubiquitin genes in higher plants. Eur. J. Biochem. 158: 57-62

Gausing K and Jensen CB (1990) Two ubiquitin-long tail fusion genes arranged as closely spaced direct repeats in barley. Gene, *in press*

Hariharan N, Kelley DE and Perry RP (1989) Equipotent mouse ribosomal protein promoters have a similar architecture that includes internal sequence elements. Genes & Development 3: 1789-1800

Hatakeyama T, Hatakeyama T and Kimura M (1988) The primary structure of ribosomal proteins L16, L23 and L33 from the archaebacterium *Halobacterium marismortui.* FEBS Letts. 240: 21-28

Hiratsuka J, Shimada H, Whittier R, Ishibashi T, Sakamoto M, Mori M, Kondo C, Honji Y, Sun CR, Meng BY, Li YQ, Kanno A, Nishizawa Y, Hirai A, Shinozaki K and Sugiura M (1989) The complete sequence of the rice (*Oryza sativa*) chloroplast genome: intermolecular recombination between distinct tRNA genes accounts for a major plastid DNA inversion during the evolution of cereals. Mol Gen Genet 217: 249-257

Joshi CP (1987) An inspection of the domain between putative TATA box and translation start site in 79 plant genes. Nucleic Acids Res. 15: 6643-6653

Köpke AKE and Wittmann-Liebold B. (1989) Comparative studies of riboso mal proteins and their genes from *Methanococcus vannielii* and other organisms. Can. J. Microbiol. 35: 11-20

Larkin JC,Hunsperger JP, Culley D, Rubenstein I and Silflow CD (1989) The organization and expression of a maize ribosomal protein gene family. Genes & Development 3: 500-509

Lund PK, Moats-Staats BM, Simmons JG, Hoyt E, D´ Ercole AJ, Martin F and Van Wyk JJ (1985) Nucleotide sequence analysis of a cDNA encoding human ubiquitin reveals that ubiquitin is synthesized as a precusor. J. Biol. Chem. 260: 7609-7613

Madsen LH, Kreiberg JD and Gausing K. A small gene family in barley encodes ribosomal proteins homologous to yeast YL17 and L22 from archaebacteria and eubacteria (*submitted*)

Mankin AS (1989) The nucleotide sequence of the genes coding for the S19 and L22 equivalent ribosomal proteins from *Halobacterium halobium*. FEBS Letts. 246: 13-16

Müntz K (1982) Seed development. In Pirson A and Zimmermann MH (eds) Encyclopedia of Plant Physiology. New Series, Vol 14A Springer-Verlag Berlin Heidelberg New York pp 505-558

Ohkubo S, Muto A, Kawauchi Y, Yamao F and Osawa S (1987) The ribosomal protein gene cluster of *Mycoplasma capricolum*. Mol Gen Genet 210: 314-322

Ohyama K, Fukuzawa H, Kohchi T, Shirai H, Sano T, Sano S, Umesono K, Shiki Y, Takeuchi M, Chang Z, Aota S-I, Inokuchi H and Ozeki H (1986) The complete nucleotide sequence of liverwort *Marchantia polymorpha* chloroplast DNA. Plant Mol. Biol. Reporter 4:148-175

Otaka E, Higo KI and Itoh T (1984) Yeast ribosomal proteins VIII. Isolation of two proteins and sequence characterization of twenty-four proteins from cytoplasmic ribosomes. Mol Gen Genet 195: 544-546

Özkaynak E, Finley D, Solomon MJ and Varshavsky A (1987) The yeast ubiquitin genes: a family of natural gene fusions. EMBO J. 6: 1429-1439

Planta RJ and Raué HA (1988) Control of ribosome biogenesis in yeast Trends Genet. 4: 64-68

Redman KL and Rechsteiner M (1989) Identification of the long ubiquitin extension as ribosomal protein S27a. Nature 338: 438-440

Tanaka M, Wakasugi T, Sugita M and Shinozaki K (1986) Genes for the eight ribosomal proteins are clustered on the chloroplast genome of tobacco (*Nicotiana tabacum*): Similarity to the S10 and *spc* operons of *Escherichia coli*. Proc. Natl. Acad. Sci. USA 83: 6030-6034

Vincentz M and Flavell R B (1989) Mapping of ribosomal RNA transcripts in wheat. The Plant Cell 1: 579-589

Wittmann-Liebold B and Greuer B (1980) Amino acid sequence of protein L22 from the large subunit of the *Escherichia coli* ribosome. FEBS Lett. 121: 105-112

Zhou D-X, Quigley F, Massenet O, and Mache R (1989) Cotranscription of the S10- and *spc*-like operons in spinach chloroplasts and identification of three of their gene products. Mol. Gen. Genet. 216: 439-445

MOLECULAR GENETICS OF CHLOROPLAST RIBOSOMES IN *CHLAMYDOMONAS*

N.W. Gillham[1], E.H. Harris[2], B.L. Randolph-Anderson[2], J.E. Boynton[2], C.R. Hauser[2], K.B. McElwain[1], and S.M. Newman[2]
Departments of Zoology[1] and Botany[2]
Duke University
Durham, NC 27706
USA

The green alga *Chlamydomonas reinhardtii* is an ideal model organism in which to investigate the cooperation of nuclear and chloroplast genomes in chloroplast ribosome formation from both developmental and evolutionary perspectives. Mutations affecting a variety of ribosomal components have been isolated and mapped genetically (see Boynton et al., 1990a and Harris, 1989 for reviews), and both the chloroplast (Boynton et al., 1988, 1990b; Blowers et al., 1989) and nuclear genomes (Debuchy et al., 1989; Kindle et al., 1989; Mayfield and Kindle, 1990; Diener et al., 1990) can be manipulated using newly developed transformation methods. In this paper we review our recent work on 1) the genetics and molecular biology of mutations to antibiotic resistance affecting the rRNA genes; 2) identification of chloroplast and nuclear genes encoding chloroplast ribosomal proteins (r-proteins); 3) expression of chloroplast genes encoding r-proteins; and 4) the transformation of chloroplast rRNA genes. We also outline the direction our current research is taking to define more precisely the rRNA-r-protein interactions involved in chloroplast ribosome biogenesis and function, as well as the molecular mechanisms involved in regulating expression of chloroplast r-protein genes.

A very high degree of conservation of sequence and secondary structure is seen between 16S and 23S rRNA genes of bacteria and of chloroplast genomes of algae and land plants (see Gutell et al., 1985; Raué et al., 1988). Mutations in these diverse organisms with similar antibiotic resistance phenotypes frequently result from identical base pair changes within highly conserved groups of bases in the same domain of the 16S or 23S rRNA molecules. Considerable evolutionary conservation is also seen for the chloroplast genes that encode a subset of the chloroplast r-proteins (Shinozaki et al., 1986; Ohyama et al., 1986; Hiratsuka et al., 1989). In most cases the corresponding *E. coli* r-protein operon can be identified in the chloroplast genome (e.g. Christopher et al., 1988; Tanaka et al., 1986; Zhou et al., 1989), with certain r-protein genes missing which have presumably migrated to the nucleus. As other papers presented in this volume show, progress is also being made in higher plants towards identification of these nuclear genes encoding chloroplast r-proteins, many of which also have homology to *E. coli* r-protein genes.

NATO ASI Series, Vol. H 55
The Translational Apparatus of Photosynthetic Organelles
Edited by R. Mache et al.

Antibiotic resistance mutations in the chloroplast rRNA genes

Antibiotic resistance mutations historically have been the most useful markers for studies of chloroplast genetics in *Chlamydomonas* (see Harris, 1989 for review). Representative mutations at seven chloroplast genetic loci defined by recombination have now been shown to result from single base-pair changes in conserved regions of the genes encoding the 16S and 23S chloroplast ribosomal RNAs (Harris et al., 1989). The order and spacing of these mutations in the physical map of the chloroplast genome are entirely consistent with the linear genetic map of the same loci (Harris et al., 1977).

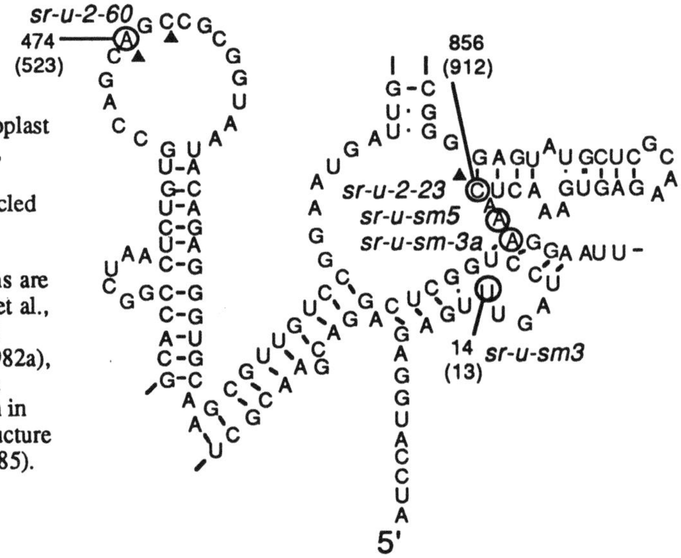

Figure 1. Part of the chloroplast 16S rRNA of *C. reinhardtii*, showing mutations to streptomycin resistance (circled bases). Positions of streptomycin resistance mutations in other organisms are marked with ▲ (see Harris et al., 1989). Bases are numbered according to Dron et al. (1982a), with the corresponding base number from *E. coli* shown in parenthesis. Secondary structure is based on Gutell et al. (1985).

In the chloroplast genome of *C. reinhardtii*, streptomycin resistance can result from changes in three different regions of the 16S rRNA gene (Figure 1; Harris et al., 1989), as well as from mutations in the *rps12* gene (Liu et al., 1989a; see below). The *sr-u-sm3* mutation, resulting in a U to G change at the nucleotide equivalent to *E. coli* 13, near the 5' end of the molecule, produces intermediate level resistance. No equivalent mutations have been reported in other organisms thus far. However, in the 16S secondary structure (Gutell et al., 1985), this region of the molecule is part of a helix in close proximity to bases mutable to streptomycin resistance in *E. coli* (Montandon et al., 1986), *Euglena* (Montandon et al., 1985) and tobacco (Etzold et al., 1987). In *C. reinhardtii*, three different base pair changes in the region of the 16S gene which corresponds to *E. coli* bases 912-915 result in mutants with low, intermediate, and high resistance to streptomycin (C to U at the position equivalent to 912, A to G at 915, and A to C at 914, respectively; Figure 1; Harris et al., 1989). Whether the level of resistance is a function only of the nucleotide affected or also depends on the base to which a given nucleotide has mutated, has not yet been established. Studies of mutations affecting the 3' terminal helix of the 16S rRNA in *E. coli* (De Stasio and Dahlberg, 1990) and the stem-loop structure implicated in spectinomycin resistance in tobacco (Fromm et al., 1987; see below), suggest that changes that disrupt base pairing often confer antibiotic resistance.

The *sr-u-2-60* mutation produces an A to C change in the "530 loop" of the 16S rRNA molecule, at the position equivalent to *E. coli* 523. Identical mutations have been found in *C. eugametos* (Gauthier et al., 1988) and *E. coli* (Melançon et al., 1988), and a C to U mutation at 525 confers streptomycin resistance in tobacco (Fromm et al., 1989). Studies of rRNA secondary and tertiary structure and r-protein cross-linking in *E. coli* suggest that all three regions implicated in streptomycin resistance may be part of a common ribosomal neighborhood that interacts with r-proteins S4, S5 and S12 (Gutell et al., 1985; Stem et al., 1988). Cognates of these proteins have been identified in chloroplast ribosomes, and the chloroplast-encoded S12 protein is highly conserved (see below).

Three different changes within the conserved region of the 16S rRNA corresponding to *E. coli* bases 1191-1193 confer varying levels of spectinomycin resistance in *C. reinhardtii*. Changing the A at 1191 to G or C produces high level resistance, whereas a G to A change at 1193 confers resistance only to lower levels of spectinomycin. Equivalent mutants have been isolated in tobacco (1193 G to A, Fromm et al., 1987). Mutation of *E. coli* 1192 from C to U or G produces high level resistance, whereas changing this base to A produces low level resistance (Sigmund et al., 1984; Makosky and Dahlberg, 1987). A mutation at the nucleotide equivalent to *E. coli* 1064, which normally pairs with 1192 within a stem region, also produces spectinomycin resistance in tobacco (Fromm et al., 1987).

Resistance to neamine and kanamycin in *C. reinhardtii* results from an A to G change at the base equivalent to *E. coli* 1408 and from a C to U change at 1409 (Harris et al., 1989). Direct action of aminoglycoside antibiotics with *E. coli* base 1408 has been demonstrated by Moazed and Noller (1987), and De Stasio and Dahlberg (1990) have obtained resistance mutations by site-directed mutagenesis of 1409 (C) and 1491 (G) which are paired at the base of a long stem region. These workers made 11 of 15 possible single and double mutations at positions 1409 and 1491 in a *rrnB* operon carried on a repressible plasmid. Changes which added bulky purine residues were lethal when expressed on the plasmid, whereas changes to C:U, U:C, or C:C, which disrupt base pairing of 1409:1491, slowed growth and conferred resistance to aminoglycoside antibiotics. The change seen in one neamine/kanamycin resistant mutant of *C. reinhardtii* (C to U at 1409) when made in *E. coli* produced only slight impairment of growth rate, coupled with low-level resistance to paromomycin but not to neamine (De Stasio and Dahlberg, 1990).

Mitochondrial mutations conferring resistance to erythromycin and to chloramphenicol have previously been mapped in yeast to three loci (*rib1 - rib3*) in the peptidyl transferase loop of the large subunit rRNA (Sor and Fukuhara, 1982, 1984). These same loci have been identified in the chloroplast gene encoding the 23S rRNA in *C. reinhardtii* (Figure 2).

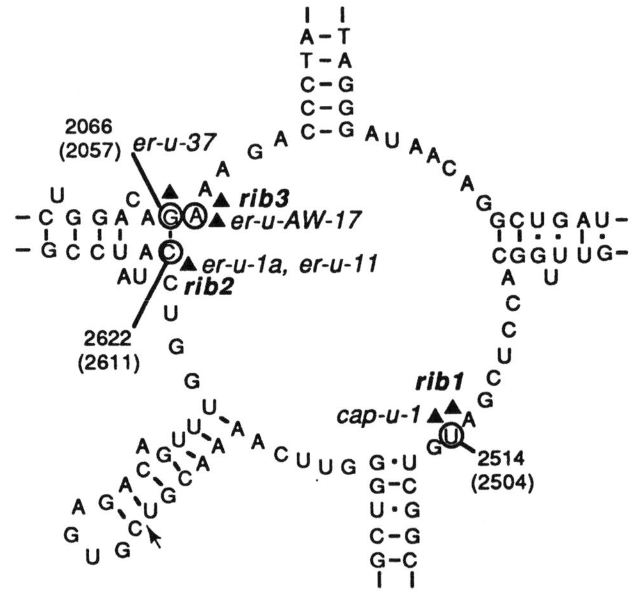

Figure 2. Peptidyl transferase region of 23S chloroplast rRNA of *C. reinhardtii*. Sites of mutations to erythromycin and chloramphenicol resistance in *C. reinhardtii* are circled; sites of corresponding mutations in other organisms are marked with ▲, and the *rib1*, *rib2* and *rib3* loci of yeast are identified (see Harris et al., 1989). Bases are numbered according to Lemieux et al. (1989), with the corresponding *E. coli* base number in parentheses. The arrow indicates the position of the 888 bp intron in the *C. reinhardtii* 23S gene (Rochaix et al., 1985).

Erythromycin resistance mutations occur at *rib3* (G to A at the base equivalent to *E. coli* 2057, A to G at 2058; Harris et al., 1989) and *rib2* (C to U or C to G at 2611; Harris et al., 1989), and a chloramphenicol resistance mutation maps at *rib1* (U to G at 2504; Bennoun et al., 1981; Bennoun, Burkhart, Harris, Boynton and Gillham, in preparation). The five erythromycin resistance mutants analyzed to date differ in their levels of cross-resistance to lincomycin and clindamycin (Harris et al., 1989; Bartlett et al., 1979). An erythromycin resistance mutation at the *rib2* locus has also been found in *C. moewusii* by Gauthier et al. (1988), and lincomycin resistance mutations at the *rib3* locus have been identified in the chloroplast genome of *Nicotiana plumbaginifolia* (Cseplö et al., 1988).

Genes for chloroplast ribosomal proteins

Five or six of the 33 r-proteins identified in the large subunit and 14 of the 31 r-proteins of the small subunit are made in the chloroplast of *C. reinhardtii*, with the remaining proteins being synthesized in the cytoplasm (Schmidt et al., 1983). The five cytoplasmically synthesized r-proteins which have been examined are all made *in vitro* as higher molecular weight precursors that can be processed by a high-speed supernatant fraction from *Chlamydomonas*, which presumably contains the stromal processing protease (Keegstra et al., 1989), to the mature form seen *in vivo* (Schmidt et al., 1984a). Antibodies to individual chloroplast r-proteins from *C. reinhardtii* have been found to crossreact with specific r-proteins from *E. coli* and with r-proteins from *Anabaena* and the chloroplast of spinach (Figure 3; Randolph-Anderson et al., 1989; Schmidt et al., 1984b). Chloroplast-synthesized r-proteins

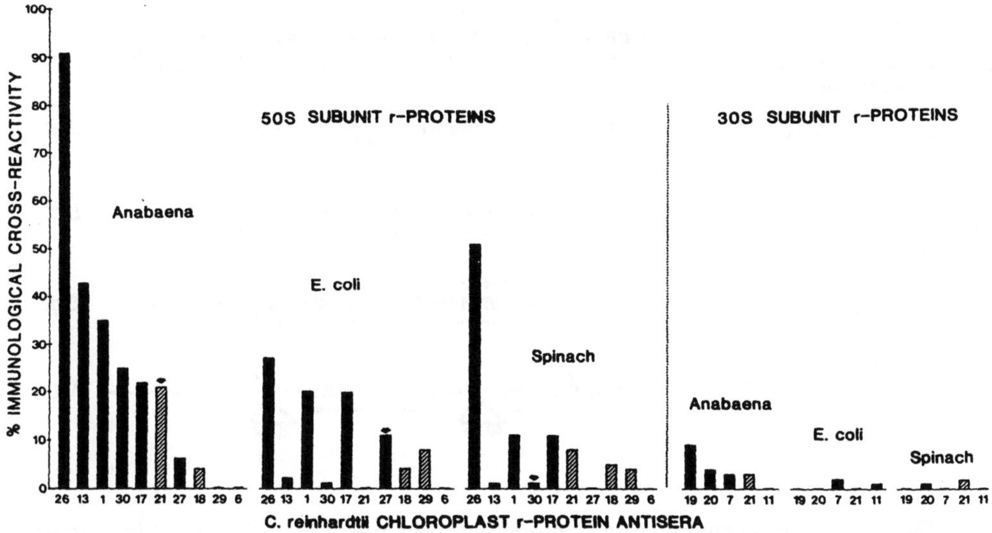

Figure 3. Histograms summarizing the immunological cross-reactivity of antiserum to specific *C. reinhardtii* chloroplast r-proteins with chloroplast r-proteins from spinach and r-proteins from *Anabaena* and *E. coli.* Antisera made against *C. reinhardtii* r-proteins synthesized in the chloroplast are shown as solid bars and those synthesized in the cytoplasm as hatched bars. Solid arrows indicate cross-reactivity seen on immunoblots of 1-D SDS gels. In cases where antiserum reacts with more than one protein, only the primary cross-reactivity is shown. From Randolph-Anderson et al. (1989).

from *C. reinhardtii* appeared to be more conserved immunologically than cytoplasmically-synthesized r-proteins, and large subunit proteins were more conserved than small subunit proteins. Among the more highly conserved large subunit r-proteins are several that are functionally related, for example r-proteins required for peptidyl transferase activity in *E. coli.*

Several chloroplast genes encoding r-proteins previously found in the completely sequenced chloroplast genomes of *Marchantia polymorpha* (Ohyama et al., 1986), tobacco (*Nicotiana tabacum*; Shinozaki et al., 1986) and rice (*Oryza sativa*; Hiratsuka et al., 1989) have also been sequenced in *C. reinhardtii.* These include *rps3* (S. Surzycki and S. Fong, pers. comm.; see Boynton et al., 1990a); *rps12* (Liu et al., 1989a); *rps19* and *rpl2* (X.-Q. Liu, pers. comm.; see Boynton et al., 1990a); *rpl14* (Lou et al., 1989) and *rpl16* (Lou et al., 1987). A chloroplast restriction fragment with homology to an *E. coli* clone containing the *rps4, rps11* and *rps13* genes has been identified by heterologous hybridization (Watson and Surzycki, 1983). Interestingly, antiserum to protein L-13 of *C. reinhardtii*, which is synthesized in the chloroplast (Schmidt et al., 1983), crossreacts with *E. coli* L5 (Schmidt et al., 1984b; Randolph-Anderson et al., 1989). While the *rpl5* gene encoding this protein is found in the chloroplast genome of *Euglena gracilis* (Christopher and Hallick, 1989) and in the cyanelle genome of *Cyanophora paradoxa* (Bryant and Stirewalt, 1990), it is absent from the completely sequenced chloroplast genomes of *Marchantia*, rice and tobacco.

Figure 4. Organization of genes equivalent to the *E. coli* *str* operon in *C. reinhardtii* (*C.r.*), *C. moewusii* (*C.m.*), *Escherichia coli* (*E.c.*), *Euglena gracilis* (*E.g.*), and *Nicotiana tabacum* (*N.t.*). Genes are identified as follows: *rps12*, r-protein S12; *rps7*, r-protein S7; *psaA*-2, exon 2 of gene encoding 84 kDa protein of Photosystem I; *psbD*, 32kDa D2 protein of Photosystem II; *tufA*, elongation factor EF-Tu; *fusA*, elongation factor G; IR, inverted repeat region. From Harris et al. (1990), modified from Liu et al. (1989a).

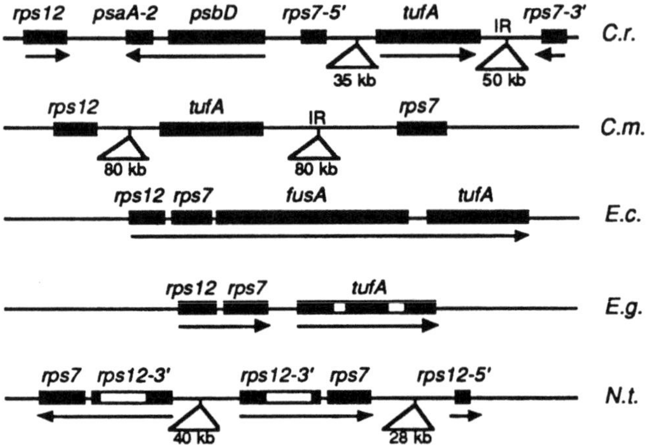

The *rps12* gene is continuous in *C. reinhardtii* (Figure 4; Liu et al., 1989a) in contrast to higher plants where the mRNA is assembled by trans-splicing transcripts from three widely separated exons (Zaita et al., 1987; Hildebrand et al., 1988). Mutations to streptomycin resistance and dependence mapping in the *sr-u-sm2* locus alter two different conserved sites in the *rps12* gene (Liu et al., 1989a) with amino acid changes identical to those in resistant and dependent mutants of *E. coli* (Funatsu and Wittmann, 1972; van Acken, 1975). A streptomycin resistance mutation in the *rps12* gene of tobacco has also been reported by Galili et al. (1989). The cloned *rps12* gene from *C. reinhardtii* is expressed in *E. coli* and its protein product, which has 68% amino acid identity with the S12 protein of *E. coli*, assembles and functions in *E. coli* ribosomes *in vivo* (Liu et al., 1989a).

In contrast to higher plants, *Euglena* and *E. coli*, the *rps7* gene of *C. reinhardtii* is not cotranscribed with *rps12* (Figure 4). Instead, the gene is split into two segments. The 5' portion of *rps7*, located near the *rps12* gene (Schmidt et al., 1985a), is separated from this gene by exon 2 of the *psaA* gene and the *psbD* gene, which are cotranscribed in the opposite direction from *rps12* (Choquet et al., 1988). The 3' portion of the *rps7* gene is located in the opposite unique sequence region and is cotranscribed with the *atpE* gene (Robertson et al., 1990). A probe to the putative 5' portion of the *rps7* gene does not hybridize to this 3'-*rps7*-*atpE* transcript. A Shine-Dalgarno sequence occurs upstream of the 3' portion of the *rps7* gene in this dicistronic mRNA, but consensus sequences for group I and II introns are absent. These results suggest that trans-splicing is probably not involved in *rps7* gene expression, but rather that the protein may be made in two pieces.

So far no nuclear genes encoding chloroplast r-proteins have been isolated from *C. reinhardtii*, but the cytoplasmically synthesized chloroplast r-proteins L-18, L-23 and L-29 (as defined by Schmidt et al., 1983) have been shown to be related to specific *E. coli* r-proteins.

Antiserum to L-18 crossreacts with *E. coli* L27 (Randolph-Anderson et al., 1989; Schmidt et al., 1984b) and L-18 has also been found to be homologous with L27 by N-terminal amino acid sequencing (Liu et al., 1988). In *C. reinhardtii* L-18 is made as a precursor which is processed in two steps, the second of which requires chloroplast protein synthesis (Schmidt et al., 1985b). The latter processing step appears to occur during ribosome assembly and may be required for maturation of the 50S ribosome subunit (Liu et al., 1988). A nuclear cDNA clone encoding the *rpl27* gene has been isolated from *Nicotiana* and sequenced by Bourque et al. (this volume).

Antiserum to *C. reinhardtii* protein L-23 crossreacts with *E. coli* L11, a protein of the GTPase center of the large subunit. A nuclear mutation of *C. reinhardtii* conferring resistance to thiostrepton (*tsp-1*) causes the loss of protein L-23 from the large subunit of the chloroplast ribosome, although L-23 continues to be synthesized. This mutant grows more slowly than wild type and its chloroplast ribosome large subunits exhibit greatly reduced activity *in vitro* (McElwain et al., in preparation). Similar thiostrepton resistance mutations in *Bacillus megaterium* and *B. subtilis* cause the loss of protein L11 from the large subunit (see Cundliffe, 1986). Protein L11 greatly enhances binding of thiostrepton to a specific site on 23S rRNA, resulting in inhibition of GTPase center function (Cundliffe, 1987). A nuclear cDNA clone isolated from spinach contains a reading frame with sequence homology to the *E. coli* gene for L11 (Subramanian et al., this volume).

Many nuclear mutations of *C. reinhardtii* are known that confer antibiotic resistance or assembly defects on chloroplast ribosomes (see Harris, 1989, for a review), but only in a few cases is there some understanding of the primary alterations. Two allelic nuclear mutations, *cr 6* and *cr-7*, cause loss from the large subunit of the cytoplasmically-synthesized r-protein L-29 and the chloroplast-synthesized L-13 (Myers et al., 1984). The primary defect in these mutants appears to be their inability to make L-29, which prevents assembly of L-13 into the 50S subunit. Antiserum to *C. reinhardtii* protein L-29 crossreacts with *E. coli* L23 and to a lesser extent with L7/L12 (Randolph-Anderson et al., 1989). Erythromycin resistance mutations at the *ery-M1* locus alter the molecular weight or electrophoretic mobility of protein L-6 of the large subunit (Davidson et al., 1974, 1978), but the gene encoding this protein has not been isolated. Antiserum directed against L-6 does not crossreact with any *E. coli* r-protein (Randolph-Anderson et al., 1989; Schmidt et al., 1984b) or with r-proteins from *Anabaena* or the chloroplast of spinach (Randolph-Anderson et al., 1989). The assembly defective mutant *ac-20* synthesizes only trace amounts of the large and small chloroplast ribosomal subunits and appears to be blocked in the processing of the 23S rRNA (Herrin and Schmidt, 1989).

Expression of r-protein mRNAs in *Chlamydomonas*

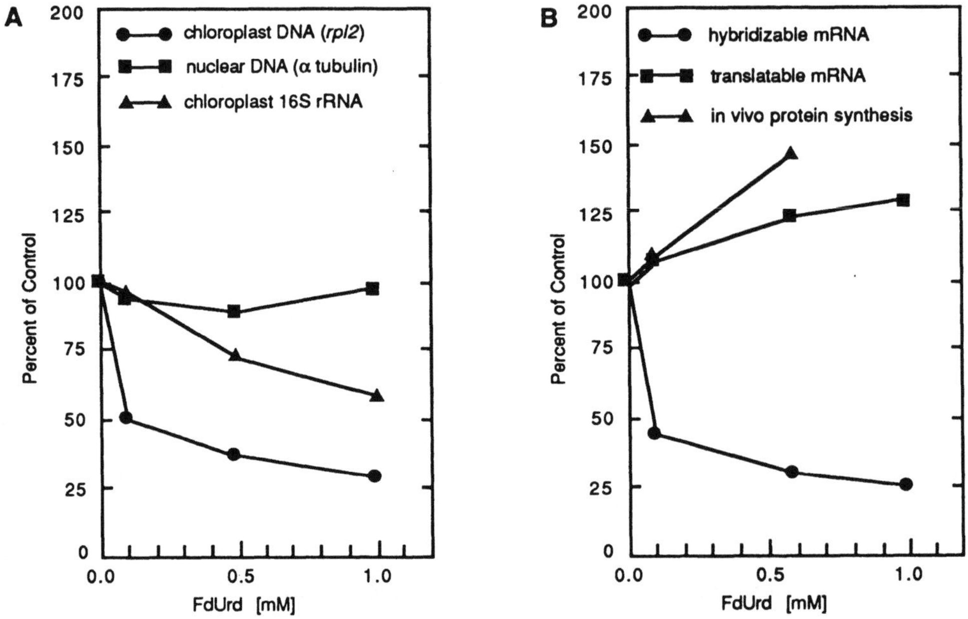

Figure 5. Effects of growth of cells in FdUrd for 7-8 generations on (A) levels of chloroplast and nuclear DNA and 16S chloroplast rRNA, and (B) on accumulation of hybridizable and translatable mRNA and rate of synthesis for chloroplast r-protein L-1 (*rpl2* gene). Data from Hosler et al. (1989).

Growth of *C. reinhardtii* in the presence of the thymidine analog 5-fluorodeoxyuridine (FdUrd) results in reduction of the ca. 80 copies of the chloroplast genome per cell (Harris, 1989) to 10-20% of normal with no effect on nuclear DNA amounts or photosynthetic growth rate (Figure 5A; Wurtz et al., 1977). Under these conditions, mRNAs for the chloroplast genes encoding ribosomal protein L-1 (*rpl2*), the α subunit of the ATP synthase (*atpA*) and the large subunit of ribulose-1,5-bisphosphate carboxylase (*rbcL*) decline with gene copy number, but the levels of translatable mRNA assayed *in vitro* in the rabbit reticulocyte system and the accumulation and rates of synthesis of these proteins *in vivo* are not reduced (Figure 5B; Hosler et al., 1989). These results suggest that mechanisms exist which mark for translation a subset of chloroplast mRNAs from genes encoding both photosynthetic proteins and chloroplast r-proteins. While the molecular mechanisms involved have not been elucidated, one can conclude that intron splicing is not required, since no introns have been detected in any of these genes in *C. reinhardtii* (Dron et al., 1982b; Hallick, 1984; X.-Q. Liu, unpublished). Whatever means is used in *C. reinhardtii* to mark chloroplast mRNAs for translation, it appears to be sufficiently general so as to be recognized by rabbit reticulocyte lysates *in vitro*, since in FdUrd-grown cells levels of protein synthesis are not reduced despite drastic declines in mRNA levels *in vitro* and *in vivo* (Figure 5). Chloroplast rRNA levels appear to decline less precipitously than those of either chloroplast DNA or mRNA.

Parallel experiments in which the amount of mitochondrial mRNA has been reduced in animal cells in tissue culture have yielded similar results (Lansman and Clayton, 1975). Photodamage of mitochondrial DNA substituted with 5-BrDU caused extensive inhibition of mitochondrial RNA synthesis, but the rate of mitochondrial protein synthesis remained unchanged for at least 48 hours after RNA synthesis had been suppressed 85%. Both sets of experiments suggest that transcription of organelle genes is constitutive and not rate limiting.

Another level of post-transcriptional regulation of chloroplast gene expression seen in *Chlamydomonas* involves the ability of chloroplast ribosomes to discriminate between mRNAs for photosynthetic and r-proteins under certain conditions. Chloroplast r-proteins are synthesized preferentially under conditions of reduced chloroplast protein synthesis (Liu et al., 1989b), while the levels of both photosynthetic and r-protein mRNAs are unaffected. This has been shown for two mutants of *C. reinhardtii*, one deficient in the small subunit of the chloroplast ribosome and the other exhibiting reduced chloroplast protein synthesis when grown in the presence of spectinomycin. Both mutants exhibit a characteristic syndrome of photosynthetic defects resulting from reduced chloroplast protein synthesis (see Harris, 1989), yet accumulate chloroplast ribosome large subunits or ribosome monomers containing a normal complement of r-proteins. Pulse-labeled cells of these mutants synthesized proteins L-1 and L-13 of the large subunit of the chloroplast ribosome at about 60% of the wild type rate, whereas the large subunit of ribulose-1,5-bisphosphate carboxylase and the α subunit of the ATP synthase were made at only 4 to 8% of the wild type rate (Table 1A). Turnover of the r-proteins and the large subunit was similar during a subsequent chase. Differences between the mutants and wild-type cells in the relative synthesis rates of these proteins were not reflected in changes in either hybridizable or translatable mRNA (Table 1B).

Why should mechanisms exist that ensure preferential translation of chloroplast r-protein mRNAs? An obvious reason is that plastid protein synthesis is necessary to sustain expression of the plastid-encoded genes whose products are required for functioning of the photosynthetic complexes. As pointed out by Walbot and Coe (1979) in explaining the plastid ribosome-deficient phenotype of the *iojap* mutation of maize, failure to synthesize chloroplast-encoded r-proteins at a sufficient rate will lead to the irreversible loss of plastid ribosomes and, therefore, plastid protein synthesis.

Inverted repeat sequences capable of forming stable stem-loop structures are ubiquitous in the 3'-untranslated region of chloroplast mRNAs (Stern and Gruissem, 1987). Binding of specific proteins to these stem-loop structures has been demonstrated in higher plants and is

Table 1. (A) Quantification of the relative rate of protein synthesis in pulse labeling experiments.using a nuclear mutant (*cr-1*) impaired in chloroplast protein synthesis and a chloroplast mutant resistant to low levels of spectinomycin (*spr-u-1-27-3*), grown in the presence of spectinomycin to reduce its capacity for chloroplast protein synthesis. Synthesis of r-protein L-1 (*rpl2* gene), the large subunit of ribulose bisphosphate carboxylase (LSU; *rbcL* gene) and the α subunit of ATP synthase (*atpA* gene) during a 5-minute pulse with $^{35}SO_4$ was quantified from two independent experiments by measuring the incorporation of label into each protein. The data were normalized on the basis of equal numbers of cells. The wild type values from the two experiments for L-1, LSU and α were 370-475, 21000-26700, and 1470-1850 cpm/10^7 cells, respectively. (B) Quantification of relative accumulation of mRNAs. Densitometer tracings of slot blot autoradiograms from three independent experiments were integrated by computer analysis. In both (A) and (B) the means and ranges (in parentheses) are presented as percentages of the values determined for wild-type cells.

(A) Relative Rates of Protein Synthesis		
protein	*cr-1*	*spr* + spec
L-1	58 (49-66)	62 (58-66)
LSU	4 (2-5)	7 (409)
α	5 (3-6	8 (7-9)

(B) Relative mRNA Accumulation				
mRNA	WT	*cr-1*	*spr* - spec	*spr* + spec
rpl2	100	74 (73-77)	92 (88-98)	249 (243-259)
rbcL	100	98 (93-100)	103 (90-108)	165 (152-176)
atpA	100	92 (82-98)	110 (87-123)	179 (166-180)

thought to play a key role in chloroplast mRNA stability (Stern et al., 1989). A stem-loop structure in the 5'-untranslated leader of the chloroplast *psbC* gene in *C. reinhardtii* has recently been implicated as the target site of a nuclear factor involved in the translation of the *psbC* mRNA (Rochaix et al., 1989). In addition, several nuclear mutations have been identified that affect synthesis or stability of specific chloroplast gene products (Rochaix et al., 1989; Kuchka et al., 1988, 1989). Thus nuclear genes in *Chlamydomonas* appear to control the expression of individual chloroplast genes at the post-transcriptional and translational levels, as is true of mitochondrial genes in yeast (Fox et al., 1988).

Transformation of the chloroplast rRNA genes

Successful transformation of the chloroplast genome of *Chlamydomonas* was achieved with the high velocity particle gun which delivers DNA coated microprojectiles into the chloroplast of the cell (Boynton et al., 1988). Stable transformants of the *atpB* and *psbA* genes, encoding the ATP synthase β subunit and photosystem II D1 reaction center protein respectively, have been obtained by complementing deletion or point mutations with cloned donor fragments containing the corresponding functional genes (Boynton et al., 1988, 1990b; Blowers et al., 1989). Antibiotic resistance mutations in the 16S and 23S rRNA genes have also been used as selectable markers for transformation of the rRNA genes in both *C. reinhardtii* and the interfertile strain *C. smithii* (Boynton et al., 1990b, Newman et al., 1990), which have virtually identical rRNA coding sequences, but differ by flanking restriction fragment length polymorphisms (RFLPs). Initial experiments to select for streptomycin-resistant transformants following microprojectile bombardment with the cloned 16S rRNA gene from a resistant mutant yielded high frequencies of nuclear mutations resistant to this antibiotic, and subsequent experiments involving bombardment with microprojectiles lacking DNA showed that the biolistic process itself is mutagenic (Newman et al., 1990). This problem was circumvented by selecting for transformation to spectinomycin resistance, since nuclear and chloroplast mutations conferring resistance to this antibiotic are much rarer than those producing streptomycin resistance, or by simultaneously selecting for streptomycin and spectinomycin resistance. RFLP differences between the chloroplast genomes of *C. reinhardtii* and *C. smithii* allowed identification of *bona fide* transformants and permitted determination of where the integration events occurred between the donor fragment and the recipient chloroplast genome.

Transformants resistant both to streptomycin and spectinomycin or to spectinomycin alone were obtained at frequencies of about 2×10^{-6}. They usually resulted from the complete or nearly complete homologous replacement of a region of the chloroplast genome in the recipient cell by the corresponding sequence from the donor plasmid. Insertion of the donor rRNA genes and flanking sequences into one inverted repeat was always followed by copy correction so that both copies of the inverted repeat acquired identical sequences. Using both genetic and physical markers, the sites of exchange for integration of the donor DNA into the recipient genome were characterized for both sets of experiments (Newman et al., 1990). Since the positions of the length variations responsible for each RFLP are known, we were able to calculate the observed recombination frequency for each interval between the scorable physical or genetic markers in the transformants and compare it to the predicted frequency of recombination (Figure 6). Exchange events between the homologous donor and recipient were not randomly distributed along the length of the recipient sequence, but occurred preferentially near the vector:insert junction, and at one hot spot upstream of the 16S gene.

Future directions

The recently developed chloroplast transformation system in *Chlamydomonas* (Boynton et al., 1988, 1990b; Blowers et al., 1989; Newman et al., 1990) provides a key methodology required for understanding the mechanisms involved in interaction between specific chloroplast r-proteins and 16S or 23S rRNAs, as well as mechanisms responsible for preferential translation of chloroplast r-protein mRNAs.

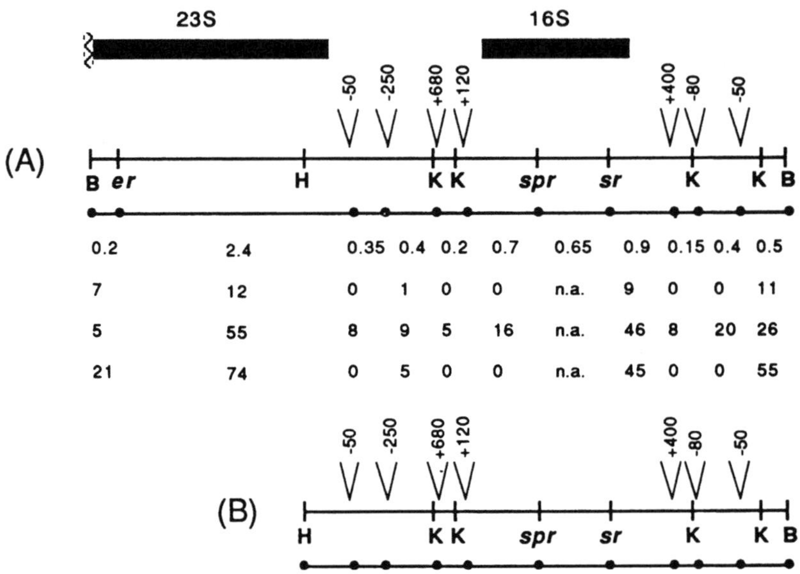

(A)

0.2	2.4	0.35	0.4	0.2	0.7	0.65	0.9	0.15	0.4	0.5	interval (kb)
7	12	0	1	0	0	n.a.	9	0	0	11	number events
5	55	8	9	5	16	n.a.	46	8	20	26	% expected/interval
21	74	0	5	0	0	n.a.	45	0	0	55	% observed/interval

(B)

0.5	0.35	0.4	0.2	0.7	0.65	0.9	0.15	0.4	0.5	interval (kb)
10	9*	0	0	1	2	6	0	0	12	number events
23	16	19	9	33	25	34	6	15	19	% expected/interval
50	45	0	0	5	10	30	0	0	60	% observed/interval

Figure 6. Comparison of expected and observed frequencies of exchange events for transformation of rRNA genes of the *C. reinhardtii* and *C. smithii* chloroplast genomes. (A) An F_1 hybrid strain (CC-1852) carrying the chloroplast genome of *C. smithii* was transformed with a pUC plasmid (P-183) containing a 7.0 kb BamHI fragment from a *C. reinhardtii* strain resistant to erythromycin, streptomycin and spectinomycin. (B) A *C. reinhardtii* strain (CC-130) resistant to streptomycin and kanamycin was transformed with a pUC plasmid (P-229) carrying a 5.3 kb BamHI-HindIII fragment from a CC-1852 derivative resistant to spectinomycin. Relative positions of restriction sites, antibiotic resistance loci and size of heterologies responsible for the RFLPs between *C. smithii* and *C. reinhardtii* are based on fine structure mapping and sequence analysis. The expected percent recombination per interval is calculated separately for the two halves of the map, from the BamHI sites to *sr* or *spr* for (A) and from *spr* to the BamHI or HindIII sites for (B), as being directly proportional to the physical distance between pairs of genetic and physical markers. Events marked by * occur within the polymorphism, generating a RFLP of intermediate length. Number and percent of events occurring between the BamHI site and *er* are derived from phenotypic analysis of 34 *spr sr* P-183/CC-1852 transformants, although the RFLP patterns of only 20 isolates were determined. Since *sr* and *spr* alleles were used as the co-selectable markers in P-183/CC-1852 transformation experiments, recombination frequencies in that interval were not applicable (n.a.). From Newman et al. (1990).

We are particularly interested in the interaction of 16S rRNA and several r-proteins involved in the codon recognition domain of the small subunit of the chloroplast ribosome. In *E. coli*, the S4, S5 and S8 proteins form a cluster with S12 on one side of the 5' domain of the small subunit (see Brimacombe, 1988) and interact in tRNA recognition to determine the efficiency and fidelity of translation (Andersson et al., 1986; Cundliffe, 1986). In both bacterial and chloroplast ribosomes, streptomycin interacts with this domain, stimulating translation and increasing misreading (Conde et al., 1975; Bartlett et al., 1979; Moazed and Noller, 1987). We are attempting site-directed mutagenesis of the *rps12* gene and the 16S rRNA gene, which are highly conserved with respect to their bacterial counterparts and in which streptomycin resistance mutations are already known, to analyze specific amino acid residues and nucleotides in 16S rRNA involved in codon recognition. We are also attempting to isolate the *rps4* and *rps5* genes of *C. reinhardtii*, and to determine whether mutations altering the function of these genes can be identified. By analogy to *E. coli* (Andersson et al., 1986; Hasenbank et al., 1973), we predict that suppressors of streptomycin dependence mutations in *rps12* may map in one of these genes. The nuclear *sr-1* locus in *C. reinhardtii* can also be mutated to confer streptomycin resistance on the small subunit of the chloroplast ribosome (Bartlett et al., 1979), although the ribosomal component altered by these mutations is as yet unknown. Mutations at a second nuclear locus, *A*, increase the resistance level of *sr-1* and chloroplast streptomycin resistant mutants. Thus the small subunit of the chloroplast ribosome in *Chlamydomonas* may be different from the small subunit of the *E. coli* ribosome, where streptomycin resistance mutations are found only in the genes encoding r-protein S12 and 16S rRNA.

We are currently exchanging the transcribed, non-translated sequences flanking the *atpB* gene, encoding the catalytic (β) subunit of the chloroplast ATPase with those of the *rps12* gene in order to define cis-acting sequences that allow the ribosome to recognize a particular class of mRNAs. Transformants can be selected by using an *atpB* deletion mutant as a recipient for the chimeric *atpB* gene (Boynton et al., 1988, 1990b). Selection for *rps12* transformants will involve use of a recipient strain possessing a mutation to streptomycin dependence in the *rps12* gene and an engineered *rps12* gene from wild type. The synthesis of the two gene products can be monitored under conditions of reduced chloroplast protein synthesis and the flanking sequences responsible for preferential translation identified. Deletion analysis and site-directed mutagenesis will then be used to define these cis-acting sequences further.

Acknowledgments

This work was supported by NIH grant GM-19427; NIH postdoctoral fellowship GM-12934 (SMN); NIH predoctoral fellowship GM-07754 (KBM); and a Hargitt postdoctoral fellowship from the Duke University Department of Zoology (CRH). We also gratefully acknowledge the technical assistance of Bryce Burkhart and Anita Johnson, and the contributions made to this research by former postdoctoral fellows J.P. Hosler, Xiang-Qin Liu, and Dominique Robertson and predoctoral students Robert Schmidt and Alan Myers.

References

Andersson DI, Andersson SGE, Kurland CG (1986) Functional interactions between mutated forms of ribosomal proteins S4, S5 and S12. Biochimie 68: 705-713

Bartlett SG, Harris EH, Grabowy CT, Gillham NW, Boynton JE (1979) Ribosomal subunits affected by antibiotic resistance mutations at seven chloroplast loci in *Chlamydomonas reinhardtii.* Mol Gen Genet 176: 199-208

Bennoun P, Delepelaire P, Delosme M (1981) A uniparental mutant of *Chlamydomonas reinhardtii* resistant to chloramphenicol. Curr Genet 3: 251-253

Blowers AD, Bogorad L, Shark KB, Sanford JC (1989) Studies on *Chlamydomonas* transformation. Foreign DNA can be stably maintained in the chromosome. Plant Cell 1: 123-132

Boynton JE, Gillham NW, Harris EH, Hosler JP, Johnson AM, Jones AR, Randolph-Anderson BL, Robertson D, Klein TM, Shark KB, Sanford JC (1988) Chloroplast transformation in *Chlamydomonas* with high velocity microprojectiles. Science 240: 1534-1538

Boynton JE, Gillham NW, Harris EH, Newman SM, Randolph-Anderson BL, Johnson AM, Jones AR (1990a) Manipulating the chloroplast genome of *Chlamydomonas* -- molecular genetics and transformation. Current Research in Photosynthesis III: 509-516

Boynton JE, Gillham NW, Newman SM, Harris EH (1990b) Organelle genetics and transformation in *Chlamydomonas.* In Hohn T (ed) Cell Organelles, Advances in Plant Gene Research, Vol 6. Springer-Verlag, in press

Brimacombe R (1988) The emerging three-dimensional structure and function of 16S ribosomal RNA. Biochemistry 27: 4207-4214

Bryant DA, Stirewalt VL (1990) The cyanelle genome of *Cyanophora paradoxa* encodes ribosomal proteins not encoded by the chloroplast genomes of higher plants. FEBS Lett 259: 273-280

Choquet Y, Goldschmidt-Clermont M, Girard-Bascou J, Kück U, Bennoun P, Rochaix J-D (1988) Mutant phenotypes support a *trans*-splicing mechanism for the expression of the tripartite *psa*A gene in the *C. reinhardtii* chloroplast. Cell 52: 903-913

Christopher DA, Hallick RB (1989) *Euglena gracilis* chloroplast ribosomal protein operon: a new chloroplast gene for ribosomal protein L5 and description of a novel organelle intron category designated group III. Nucl Acids Res 17: 7591-7608

Christopher DA, Cushman JC, Price CA, Hallick RB (1988) Organization of ribosomal protein genes *rpl23, rpl2, rps19, rpl22* and *rps3* on the *Euglena gracilis* chloroplast genome. Curr Genet 14: 275-286

Conde MF, Boynton JE, Gillham NW, Harris EH, Tingle CL, Wang WL (1975) Chloroplast genes in *Chlamydomonas* affecting organelle ribosomes. Genetic and biochemical analysis of antibiotic-resistant mutants at several gene loci. Mol Gen Genet 140: 183-220

Cseplö A, Etzold T, Schell J, Schreier PH (1988) Point mutations in the 23S rRNA genes of four lincomycin resistant *Nicotiana plumbaginifolia* mutants could provide new selectable markers for chloroplast transformation. Mol Gen Genet 214: 295-299

Cundliffe E (1986) Involvement of specific portions of ribosomal RNA in defined ribosomal functions: a study using antibiotics. In Hardesty, B, Kramer G (eds) Structure, Function, and Genetics of Ribosomes. Springer-Verlag. pp. 586-604

Cundliffe E (1987) On the nature of antibiotic binding sites in ribosomes. Biochimie 69: 863-869

Davidson JN, Hanson MR, Bogorad L (1974) An altered chloroplast ribosomal protein in *ery-M1* mutants of *Chlamydomonas reinhardi.* Mol Gen Genet 132: 119-129

Davidson JN, Hanson MR, Bogorad L (1978) Erythromycin resistance and the chloroplast ribosome in *Chlamydomonas reinhardi.* Genetics 89: 281-297

Debuchy R, Purton S, Rochaix J-D (1989) The argininosuccinate lyase gene of *Chlamydomonas reinhardtii*: an important tool for nuclear transformation and for correlating the genetic and molecular maps of the *ARG7* locus. EMBO J 8: 2803-2809

De Stasio EA, Dahlberg AE (1990) Effects of mutagenesis of a conserved base-paired site near the decoding region of *Escherichia coli* 16S ribosomal RNA. J Mol Biol 212: 127-133

Diener DR, Curry AM, Johnson KA, Williams BD, Lefebvre PA, Kindle KL, Rosenbaum JL (1990) Rescue of a paralyzed-flagella mutant of *Chlamydomonas* by transformation. Proc Natl Acad Sci USA 87: 5739-5743

Dron M, Rahire M, Rochaix J-D (1982a) Sequence of the chloroplast DNA region of *Chlamydomonas reinhardii* containing the gene of the large subunit of ribulose bisphosphate carboxylase and parts of its flanking genes. J Mol Biol 162: 775-793

Dron M, Rahire M, Rochaix J-D (1982b) Sequence of the chloroplast 16S rRNA gene and its surrounding regions of *Chlamydomonas reinhardii*. Nucl Acids Res 10: 7609-7620

Etzold T, Fritz CC, Schell J, Schreier PH (1987) A point mutation in the chloroplast 16S rRNA gene of a streptomycin resistant *Nicotiana tabacum*. FEBS Lett 219: 343-346

Fox TD, Costanzo MC, Strick CA, Marykwas DL, Seaver EC, Rosenthal JK (1988) Translational regulation of mitochondrial gene expression by nuclear genes of *Saccharomyces cerevisiae*. Phil Trans Royal Soc London B 319: 97-105

Fromm H, Edelman M, Aviv D, Galun E (1987) The molecular basis for rRNA-dependent spectinomycin resistance in *Nicotiana* chloroplasts. EMBO J 6: 3233-3237

Fromm H, Galun E, Edelman M (1989) A novel site for streptomycin resistance in the "530 loop" of chloroplast 16S ribosomal RNA. Plant Mol Biol 12: 499-505

Funatsu G, Wittmann HG (1972) Ribosomal proteins XXXIII. Location of amino acid replacements in protein S12 isolated from *Escherichia coli* mutants resistant to streptomycin. J Mol Biol 68: 547-550

Galili S, Fromm H, Aviv D, Edelman M, Galun E (1989) Ribosomal protein S12 as a site for streptomycin resistance in *Nicotiana* chloroplasts. Mol Gen Genet 218: 289-292

Gauthier A, Turmel M, Lemieux C (1988) Mapping of chloroplast mutations conferring resistance to antibiotics in *Chlamydomonas:* Evidence for a novel site of streptomycin resistance in the small subunit rRNA. Mol Gen Genet 214: 192-197

Gutell RR, Weiser B, Woese CR, Noller HF (1985) Comparative anatomy of 16-S-like ribosomal RNA. Prog Nucl Acid Res Mol Biol 32: 155-216

Hallick RB (1984) Identification and partial DNA sequence of the gene for the α-subunit of the ATP synthase complex of *Chlamydomonas reinhardii* chloroplasts. FEBS Lett 177: 274-276

Harris EH (1989) The *Chlamydomonas* Sourcebook. Academic Press, San Diego, 780 pp.

Harris EH, Boynton JE, Gillham NW, Tingle CL, Fox SB (1977) Mapping of chloroplast genes involved in chloroplast ribosome biogenesis in *Chlamydomonas reinhardtii*. Mol Gen Genet 155: 249-265

Harris EH, Burkhart BD, Gillham NW, Boynton JE (1989) Antibiotic resistance mutations in the chloroplast 16S and 23S rRNA genes of *Chlamydomonas reinhardtii:* Correlation of genetic and physical maps of the chloroplast genome. Genetics 123: 281-292

Harris EH, Boynton JE, Gillham NW, Burkhart BD, Newman SM (1990) Chloroplast genome organization in *Chlamydomonas*. Arch Protistenk, in press

Hasenbank R, Guthrie C, Stöffler G, Wittmann HG, Rosen L, Apirion D (1973) Electrophoretic and immunological studies on ribosomal proteins of 100 *Escherichia coli* revertants from streptomycin dependence. Mol Gen Genet 127: 1-18

Herrin DL, Schmidt GW (1989) Self-splicing of the chloroplast 23S intron of *Chlamydomonas*. Plant Physiol 89: 192s

Hildebrand M, Hallick RB, Passavant CW, Bourque DP (1988) Trans-splicing in chloroplasts: The *rps12* loci of *Nicotiana tabacum*. Proc Natl Acad Sci USA 85: 372-376

Hiratsuka J, Shimada H, Whittier R, Ishibashi T, Sakamoto M, Mori M, Kondo C, Honji Y, Sun C-R, Meng B-Y, Li Y-Q, Kanno A, Nishizawa Y, Hirai A, Shinozaki K, Sugiura M (1989) The complete sequence of the rice (*Oryza sativa*) chloroplast genome: intermolecular recombination between distinct tRNA genes accounts for a major plastid DNA inversion during the evolution of the cereals. Mol Gen Genet 217: 185-194

Hosler JP, Wurtz EA, Harris EH, Gillham NW, Boynton JE (1989) The relationship between plastid gene dosage and gene expression in the chloroplast of *Chlamydomonas reinhardtii*. Plant Physiol 91: 648-655

Keegstra K, Olsen LJ, Theg SM (1989) Chloroplastic precursors and their transport across the envelope membranes. Ann Rev Plant Physiol Plant Mol Biol 40: 471-501

Kindle KL, Schnell RA, Fernández E, Lefebvre PA (1989) Stable nuclear transformation of *Chlamydomonas* using the *Chlamydomonas* gene for nitrate reductase. J Cell Biol 109: 2589-2601

Kuchka MR, Mayfield SP, Rochaix J-D (1988) Nuclear mutations specifically affect the synthesis and/or degradation of the chloroplast-encoded D2 polypeptide of photosystem II in *Chlamydomonas reinhardtii*. EMBO J 7: 319-324

Kuchka MR, Goldschmidt-Clermont M, van Dillewijn J, Rochaix J-D (1989) Mutation at the *Chlamydomonas* nuclear *NAC2* locus specifically affects stability of the chloroplast *psbD* transcript encoding polypeptide D2 of PS II. Cell 58: 869-876

Lansman RA, Clayton DA (1975) Mitochondrial protein synthesis in mouse L-cells: Effect of selective nicking of mitochondrial DNA. J Mol Biol 99: 777-793

Lemieux C, Boulanger J, Otis C, Turmel M (1989) Nucleotide sequence of the chloroplast large subunit rRNA gene from *Chlamydomonas reinhardtii*. Nucl Acids Res 17: 7997

Liu X-Q, Gillham NW, Boynton JE (1988) Chloroplast ribosomal protein L-18 in *Chlamydomonas reinhardtii* is processed during ribosome assembly. Mol Gen Genet 214: 588-591

Liu X-Q, Gillham NW, Boynton JE (1989a) Chloroplast ribosomal protein gene *rps12* of *Chlamydomonas reinhardtii*. Wild-type sequence, mutation to streptomycin resistance and dependence, and function in *Escherichia coli*. J Biol Chem 264: 16100-16108

Liu X-Q, Hosler JP, Boynton JE, Gillham NW (1989b) mRNAs for two ribosomal proteins are preferentially translated in the chloroplast of *Chlamydomonas reinhardtii* under conditions of reduced protein synthesis. Plant Mol Biol 12: 385-394

Lou JK, Wu M, Chang CH, Cuticchia AJ (1987) Localization of a r-protein gene within the chloroplast DNA replication origin of *Chlamydomonas*. Curr Genet 11: 537-541

Lou JK, Cruz FD, Wu M (1989) Nucleotide sequence of the chloroplast ribosomal protein gene L14 in *Chlamydomonas reinhardtii*. Nucl Acids Res 17: 3587

Makosky PC, Dahlberg AE (1987) Spectinomycin resistance at site 1192 in 16S ribosomal RNA of *E. coli*: an analysis of three mutants. Biochimie 69: 885-889

Mayfield SP, Kindle KL (1990) Stable nuclear transformation of *Chlamydomonas reinhardtii* by using a *C. reinhardtii* gene as the selectable marker. Proc Natl Acad Sci USA 87: 2087-2091

Melançon P, Lemieux C, Brakier-Gingras L (1988) A mutation in the 530 loop of *Escherichia coli* 16S ribosomal RNA causes resistance to streptomycin. Nucl Acids Res 16: 9631-9639

Moazed D, Noller HF (1987) Interaction of antibiotics with functional sites in 16S ribosomal RNA. Nature 327: 389-394

Montandon P-E, Nicolas P, Schürmann P, Stutz E (1985) Streptomycin-resistance of *Euglena gracilis* chloroplasts: identification of a point mutation in the 16S rRNA gene in an invariant position. Nucl Acids Res 13: 4299-4310

Montandon P-E, Wagner R, Stutz E (1986) *E. coli* ribosomes with a C912 to U base change in the 16S rRNA are streptomycin resistant. EMBO J 5: 3705-3708

Myers AM, Harris EH, Gillham NW, Boynton JE (1984) Mutations in a nuclear gene of *Chlamydomonas* cause the loss of two chloroplast ribosomal proteins, one synthesized in the chloroplast and the other in the cytoplasm. Curr Genet 8: 369-378

Newman SM, Boynton JE, Gillham NW, Randolph-Anderson BL, Johnson AM, Harris EH (1990) Transformation of chloroplast ribosomal RNA genes in *Chlamydomonas*: Molecular and genetic characterization of integration events. Genetics, in press.

Ohyama K, Fukuzawa H, Kohchi T, Shirai H, Sano T, Sano S, Umesono K, Shiki Y, Takeuchi M, Chang Z, Aota S, Inokuchi H, Ozeki H (1986) Chloroplast gene organization deduced from complete sequence of liverwort *Marchantia polymorpha* chloroplast DNA. Nature 322: 572-574

Randolph-Anderson BL, Boynton JE, Gillham NW (1989) Electrophoretic and immunological comparisons of chloroplast and prokaryotic ribosomal proteins reveal that certain families of large subunit proteins are evolutionarily conserved. J Mol Evol 29: 68-88

Raué HA, Klootwijk J, Musters W (1988) Evolutionary conservation of structure and function of high molecular weight ribosomal RNA. Prog Biophys Mol Biol 51: 77-129

Robertson D, Boynton JE, Gillham NW (1990) Cotranscription of the wild type chloroplast *atpE* gene encoding the CF_1/CF_0 epsilon subunit with the 3' half of the *rps7* gene in *Chlamydomonas reinhardtii* and characterization of frameshift mutations in *atpE*. Mol Gen Genet 221: 155-163

Rochaix J-D, Rahire M, Michel F (1985) The chloroplast ribosomal intron of *Chlamydomonas reinhardii* codes for a polypeptide related to mitochondrial maturases. Nucl Acids Res 13: 975-984

Rochaix J-D, Kuchka M, Mayfield S, Schirmer-Rahire M, Girard-Bascou J, Bennoun P (1989) Nuclear and chloroplast mutations affect the synthesis or stability of the chloroplast *psbC* gene product in *Chlamydomonas reinhardtii*. EMBO J 8: 1013-1021

Schmidt RJ, Richardson CB, Gillham NW, Boynton JE (1983) Sites of synthesis of chloroplast ribosomal proteins in *Chlamydomonas*. J Cell Biol 96: 1451-1463

Schmidt RJ, Myers AM, Gillham NW, Boynton JE (1984a) Chloroplast ribosomal proteins of *Chlamydomonas* synthesized in the cytoplasm are made as precursors. J Cell Biol 98: 2011-2018

Schmidt RJ, Myers AM, Gillham NW, Boynton JE (1984b) Immunological similarities between specific chloroplast ribosomal proteins from *Chlamydomonas reinhardtii* and ribosomal proteins from *E. coli*. Mol Biol Evol 1: 317-334

Schmidt RJ, Hosler JP, Gillham NW, Boynton, JE (1985a) Biogenesis and evolution of chloroplast ribosomes: cooperation of nuclear and chloroplast genes. In Steinback KE, Bonitz S, Arntzen CJ, Bogorad L, (eds) Molecular Biology of the Photosynthetic Apparatus, Cold Spring Harbor Laboratory, pp. 417-427

Schmidt RJ, Gillham NW, Boynton JE (1985b) Processing of the precursor to a chloroplast ribosomal protein made in the cytosol occurs in two steps, one of which depends on a protein made in the chloroplast. Mol Cell Biol 5: 1093-1099

Shinozaki K, Ohme M, Tanaka M, Wakasugi T, Hayashida N, Matsubayashi T, Zaita N, Chunwongse J, Obokata J, Yamaguchi-Shinozaki K, Ohto C, Torazawa K, Meng BY, Sugita M, Deno H, Kamogashira T, Yamada K, Kusuda J, Takaiwa F, Kato A, Tohdoh N, Shimada H, Sugiura M (1986) The complete nucleotide sequence of the tobacco chloroplast genome: its gene organization and expression. EMBO J 5: 2043-2049

Sigmund CD, Ettayebi M, Morgan EA (1984) Antibiotic resistance mutations in 16S and 23S ribosomal RNA genes of *Escherichia coli*. Nucl Acids Res 12: 4653-4663

Sor F, Fukuhara H (1982) Identification of two erythromycin resistance mutations in the mitochondrial gene coding for the large ribosomal RNA in yeast. Nucl Acids Res 10: 6571-6577

Sor F, Fukuhara H (1984) Erythromycin and spiramycin resistance mutations of yeast mitochondria: nature of the *rib2* locus in the large ribosomal RNA gene. Nucl Acids Res 12: 8313-8318

Stern DB, Gruissem W (1987) Control of plastid gene expression: 3' inverted repeats act as mRNA processing and stabilizing elements, but do not terminate transcription. Cell 51: 1145-1157

Stern DB, Jones H, Gruissem W (1989) Function of plastid mRNA 3' inverted repeats. RNA stabilization and gene-specific protein binding. J Biol Chem 264: 18742-18750

Stern S, Weiser B, Noller HF (1988) Model for the three-dimensional folding of 16 S ribosomal RNA. J Mol Biol 204: 447-481

Tanaka M, Wakasugi T, Sugita M, Shinozaki K, Sugiura M (1986) Genes for the eight ribosomal proteins are clustered on the chloroplast genome of tobacco (*Nicotiana tabacum*): Similarity to the S10 and spc operons of *Escherichia coli*. Proc Natl Acad Sci USA 83: 6030-6034

van Acken U (1975) Protein chemical studies on ribosomal proteins S4 and S12 from *ram* (ribosomal ambiguity) mutants of *Escherichia coli*. Mol Gen Genet 140: 61-68

Walbot V, Coe EH Jr (1979) Nuclear gene *iojap* conditions a programmed change to ribosome-less plastids in *Zea mays*. Proc Natl Acad Sci USA 76: 2760-2764

Watson JC, Surzycki SJ (1983) Both the chloroplast and nuclear genomes of *Chlamydomonas reinhardi* share homology with *Escherichia coli* genes for transcriptional and translational components. Curr Genet 7: 201-210

Wurtz EA, Boynton JE, Gillham NW (1977) Perturbation of chloroplast DNA amounts and chloroplast gene transmission in *Chlamydomonas* reinhardtii by 5-fluorodeoxyuridine. Proc Natl Acad Sci USA 74: 4552-4556

Zaita N, Torazawa K, Shinozaki K, Sugiura M (1987) Trans splicing in vivo: joining of transcripts from the "divided" gene for ribosomal protein S12 in the chloroplasts of tobacco. FEBS Lett 210: 153-156

Zhou D-X, Quigley F, Massenet O, Mache R (1989) Cotranscription of the S10- and spc-like operons in spinach chloroplasts and identification of three of their gene products. Mol Gen Genet 216: 439-445

Chloroplast Ribosomal Protein Operons of *Euglena gracilis*

Richard B. Hallick[1, 2], David A. Christopher[2, 3], Donald W. Copertino[2], Robert G. Drager[2], Kristin P. Jenkins[2], and Jennifer K. Stevenson[2]
Department of Biochemistry[1] and Department of Molecular and Cellular Biology[2], University of Arizona, Tucson, AZ 85721 U.S.A.

SUMMARY

The *Euglena gracilis* chloroplast genome encodes at least 17 chloroplast ribosomal proteins. Most are present in polycistronic operons. Known coding loci have the following gene organizations: (1) *rps12-rps7-tufA*, *(2) rpl20*, (3) *rpl23-rpl2-rps19-rpl22-rps3*-orf516-*rpl16-rpl14-rpl5-rps8-rpl36-trnI-rps14*, (4) *rps4-rps11*, and (5) *rps2-atpI-atpH-atpF-atpA*. These 17 chloroplast ribosomal protein genes compare in number to the 19 found in higher plant chloroplast genomes. Many of the *Euglena* genes contain group III introns, which are unique to the *Euglena* chloroplast genome. In this article, the structure, transcription, introns and RNA processing pathways of *Euglena* chloroplast ribosomal protein operons will be reviewed.

ORGANIZATION OF RIBOSOMAL PROTEIN GENES

rps12-rps7-tufA. The structural gene for protein synthesis elongation factor EF-Tu was originally mapped by heterologous hybridization with the corresponding genes from *E. coli* and *Chlamydomonas* chloroplast DNAs (Passavant et al, 1983). This *tufA* gene was cloned and sequenced by Montandon and Stutz (1983). It was found to encode a polypeptide of 408 amino acids with 70% amino acid sequence identity to the corresponding *E. coli* EF-Tu polypeptide. Quite unexpectedly, the *tufA* locus was found to be interrupted by two small introns of 103 and 110 nucleotides. Montandon and Stutz (1984) subsequently characterized a dicistronic *rps12-rps7* operon immediately upstream of *tufA*. The gene organization is *rps12*-80 bp spacer-*rps7*-174 bp spacer-*tufA*. This gene arrangement is

---

[3] Current address: Department of Biochemistry and Biophysics, Texas A and M University, College Station, Texas 77843, U.S.A.

NATO ASI Series, Vol. H 55
The Translational Apparatus of Photosynthetic Organelles
Edited by R. Mache et al.

the same as that of the *E. coli str* operon, except for the absence of the gene for EF-G between *rps7* and *tufA*. In *Euglena*, the *rps12-rps7* genes are co-transcribed, whereas the *tufA* gene is transcribed separately in a discistronic operon with a distal gene of unknown function designated orf206 (Montandon and Stutz, 1984, 1987). The untranslated leader of the *tufA*-orf206 pre-mRNA also contains a small intron (Montandon and Stutz, 1987). The location and gene organization of the *rps12-rps7-tufA* genes are shown in Figures 1 and 2.

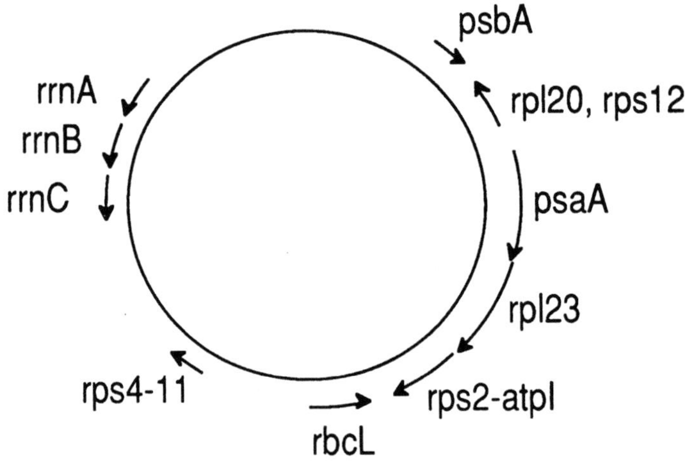

Figure 1. Circular Map of the *Euglena gracilis* chloroplast genome. The positions of the ribosomal protein coding loci relative to other known genes is shown. Arrows denote the direction of transcription.

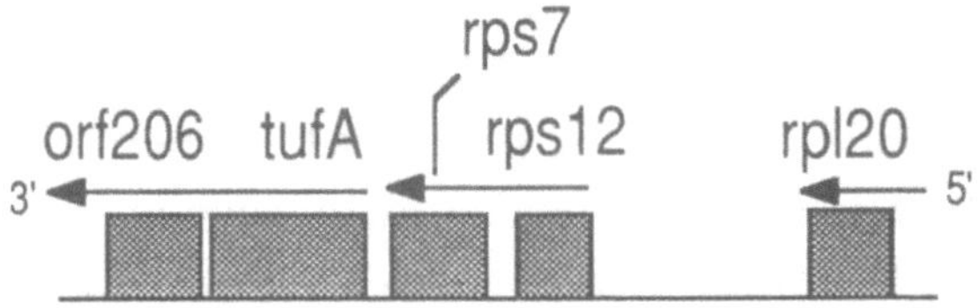

Figure 2. Organization of the *rpl20* and *rps12-rps7-tufA* genes.

rpl20. As shown in Figure 2, the gene for ribosomal protein L-20 is located 783 bp upstream from the *rps12-rps7* operon (Manzara and Hallick, 1987). To date, transcripts from the *rpl20* locus have not been characterized, and no genes have been identified in the *rpl20-rps12* intergenic spacer region.

rpl23-rpl2-rps19-rpl22-rps3-orf516-*rpl16-rpl14-rpl5-rps8-rpl36-trnI-rps14* operon. The largest *Euglena* chloroplast ribosomal protein operon encodes 11 ribosomal proteins, an internal tRNA cistron, and a large, unidentified open reading frame of 516 codons (Christopher et al, 1988; Nickoloff et al., 1989; Christopher and Hallick, 1989, 1990). The location of this operon is shown in Figure 1, and the overall organization of the genes is shown in Figure 3. This operon begins as in higher plants with the five genes *rpl23-rpl2-rps19-rpl22-rps3* (Christopher et al, 1988). The *rps3-rpl16* intercistronic region encodes an unidentified orf of 516 codons interrupted by several introns (K. P. Jenkins, D. A. Christopher, and R. B. Hallick, unpublished observations). Downstream is a second cluster of ribosomal protein genes, punctuated by one internal tRNA gene. The gene order is *rpl16-rpl14-rpl5-rps8-rpl36-trnI-rps14* (Christopher and Hallick, 1989).

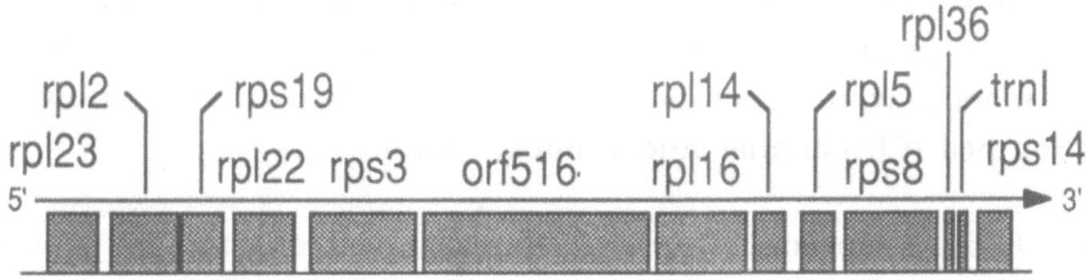

Figure 3. Organization of the *rpl23* operon. See Figure 1 for location.

There appears to be an overall evolutionary conservation of gene order for many clusters of ribosomal protein genes. The *Euglena* gene order is quite similar to that of tobacco, rice and *Marchantia polymorpha* chloroplast genomes, which are all *rpl23-rpl2-rps19-rpl22-rps3-rpl16-rpl14-rps8-infA-rpl36-rps11-rpoA*. The most unique gene of the *Euglena* operon is *rpl5*. It is not encoded in the sequenced genomes of tobacco, liverwort, or rice, nor in the partial chloroplast sequences from other species. The organization of these ribosomal protein genes is also strikingly similar to the gene arrangement in the S10 and *spc* ribosomal protein operons of *E. coli*.

Many of the ribosomal protein genes in the *rpl23* operon contain small introns, designated "group III" introns (Christopher and Hallick, 1989). The properties of the group III introns are: (i) small, uniform size, range 93-111 nt; (ii) degenerate versions of group II 5'- and 3'-intron boundary sequences; (iii) no conserved secondary structure and (iv) location primarily in genes for proteins involved in translation and transcription. The *rpl23*, *rps19*, and *rps3* loci contain 2-3 introns each. The *rpl16*, *rpl14*, *rps8*, and *rps14* genes contain 1-3 introns each. Of the 21 introns in this operon, seven (*rps3*-intron 1, *rpl16* intron 2, and *rps8* introns 1 and 3, orf516 introns 1, 3, and 4) are group II introns. The remaining 14 introns are group III.

The RNA maturation pathway for the *rpl23* operon has been studied (Christopher and Hallick, 1990). Gene-specific probes for each of the 13 genes in the operon, including the internal tRNA cistron, were all found to hybridize to a common pre-mRNA of estimated size 8.3 kb. Thus, all 13 cistrons, beginning with the *rpl23* locus and ending with *rps14*, are initially transcribed as a single, polycistronic pre-mRNA-tRNA, containing at least 21 group II and group III introns. The 8.3 kb fully spliced precursor is further processed by numerous intercistronic cleavage events. The ribosomal protein mRNAs accumulate primarily as penta-, tetra-, tri-, di-, and monocistronic species. A detailed RNA maturation pathway has been described (Christopher and Hallick, 1990).

rps4-rps11 operon. *Euglena* has a novel chloroplast ribosomal protein operon encoding ribosomal proteins S4 and SlI. It is located upstream of, and in the same polarity as a previously described leucine tRNA gene (Orozco and Hallick, 1982). In most higher plants, the *rps11* gene is co-transcribed with the *rpoA* gene, encoding an RNA polymerase subunit. The plant chloroplast *rps4* gene is apparently transcribed as a monocistronic operon. In *E. coli*, the *rps4* and *rps11* genes are within the α-operon, organized *rps13-rps11-rps4-rpoA-rpl17*. The coding region of *Euglena* chloroplast *rps11* is interrupted by two group III introns of 107 and 100 nt. The major transcript from the *Euglena* operon is a discistronic *rps4-rps11* mRNA. Monocistronic species are not detected (J. K. Stevenson and R. B. Hallick, submitted for publication). In characterizing the *rps4-rps11* operon, the polymerase chain reaction was used to synthesize a double strand cDNA copy of the spliced *rps4-rps11* mRNA. From the cDNA sequence, the two group III introns of 107 and 100 nt were confirmed within *rps11*, flanking a 13 nt exon. A third group III intron of 95 nt was identified within the *rps4-rps11* intercistronic spacer, located between the *rps11* ribosome binding site and the ATG-initiator codon. The location of this group III intron is somewhat similar to the intron location in the *tufA* nontranscribed leader (Montandon and Stutz (1987).

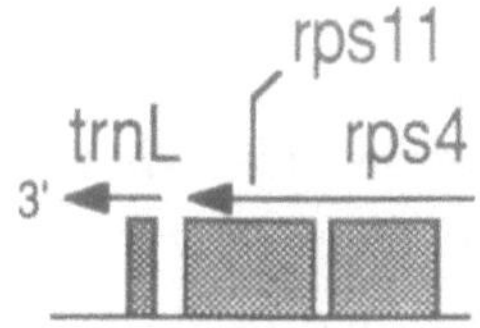

Figure 4. Organization of the rps4-rps11 operon. See Figure 1 for location.

rps2-atpI-atpF-atpH-atpA operon. As was described for spinach, pea (Hudson et al, 1987) and other chloroplast DNAs, the *Euglena* gene for chloroplast ribosomal protein S2 is encoded with four cistrons for ATP synthase subunits. The *rps2-atpI-F-H-A* operon is located just downstream, and in the same polarity as the *rpl23* operon (Figure 1), separated by two tRNA genes, *trnF* and *trnC* (Nickoloff et al., 1989). The *Euglena* gene order of *rps2-atpI-atpF-atpH-atpA* is the same as that of spinach, tobacco, liverwort, pea, wheat, and rice chloroplasts. The *Euglena rps2* gene is unique in that it has 1 group II intron and 3 group III introns. The arrangement of the genes in this operon is shown in Figure 5. All five cistrons are initially transcribed as a polycistronic pre-mRNA. RNA maturation includes splicing of the introns in *rps2* and other cistrons, and intercistronic cleavage to yield a monocistronic, fully-spliced *rps2* mRNA (R. G. Drager and R. B. Hallick, unpublished observations).

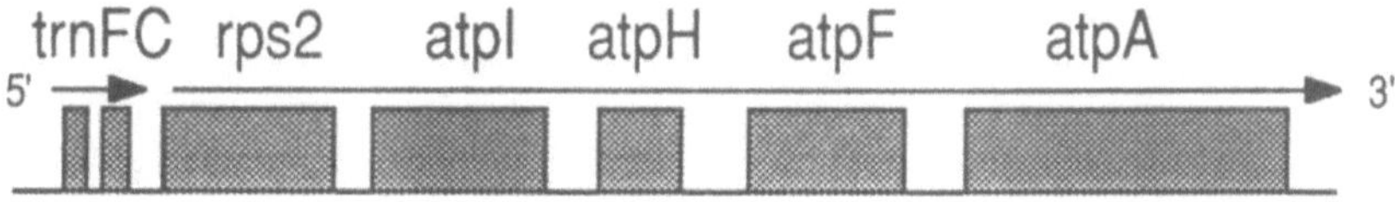

Figure 5. Organization of the *rps2-atpI-atpF-atpH-atpA* operon.

RIBOSOMAL PROTEIN GENES OF *EUGLENA* CHLOROPLASTS COMPARED TO OTHER KNOWN GENOMES

The distribution of chloroplast ribosomal protein genes between the chloroplast and the nucleus in photosynthetic eukaryotes has been extensively conserved throughout evolution. Differences in ribosomal protein coding capacity among chloroplast genomes exist for only a few genes. Both *Euglena* and *Chlamydomonas* have a chloroplast *tufA* locus, which is absent in known higher plant plastids. A list of all known ribosomal protein genes encoded by tobacco, liverwort, rice, and *Euglena* chloroplast genomes is presented in Table 1.

Table 1. The ribosomal protein genes known to be encoded in either the *Euglena*, tobacco, liverwort, or rice chloroplast genomes. A plus (+) signifies a known gene, a minus (-) means gene is known to be absent, and a question mark (?) denotes coding location is not yet determined. An asterisk (*) denotes an intron-containing gene.

	Chloroplast Genome			
Gene	Euglena	Tobacco	Liverwort	Rice
rps2	+*	+	+	+
rps3	+*	+	+	+
rps4	+	+	+	+
rps7	+	+	+	+
rps8	+*	+	+	+
rps11	+*	+	+	+
rps12	+	+*	+*	+*
rps14	+*	+	+	+
rps15	?	+	+	+
rps16	?	+*	-	+*
rps18	?	+	+	+
rps19	+*	+	+	+
rpl2	+	+*	+*	+*
rpl5	+	-	-	-
rpl14	+*	+	+	+
rpl16	+*	+*	+*	+*
rpl20	+	+	+	+
rpl21	?	-	+	-
rpl22	+	+	+	+
rpl23	+*	+	+	+
rpl33	?	+	+	+
rpl36	+	+	+	+
tufA	+*	-	-	-
infA	?	+	+	+

The tobacco and rice chloroplast genomes differ from liverwort in only one of their 20 ribosomal protein genes, with *rpl21* being present only in liverwort and *rps16* only in tobacco and rice. Therefore, the presence of a new ribosomal protein gene, *rpl5*, in *Euglena* is noteworthy. Other ribosomal protein genes not yet found in *Euglena*, but present in at least one of the higher plant chloroplast genomes are *rps15*, *rps18*, and *rpl33*. The possibility exists that one or more of these genes will be identified in the *Euglena* chloroplast genome, since approximately 30-40 kbp of the 145 kbp genome has not yet been sequenced.

INTRONS OF *EUGLENA* CHLOROPLAST RIBOSOMAL PROTEIN GENES

Without doubt, the most novel feature of the *Euglena* chloroplast ribosomal protein genes are their introns. A list of 25 introns in the various genes is presented in Table 2. The introns fall into two major categories that can be distinguished by size. The introns in the size range 277-356 are of the category group II as described in the review by Michel et al (1989). The most distinguishing feature of these introns is the secondary structure of domains V and VI (Christopher et al, 1989). Other features characteristic of group II domains I-IV are also evident.

Table 2. Sizes of Introns of *Euglena* Chloroplast Ribosomal RNA Genes.

Gene	No. of Introns	Intron Size(s)
rpl14	1	108
rpl16	3	98, 356, 208
rpl22	1	347
rpl23	3	106, 99, 103
rps11	2	107, 100
rps14	1	106
rps19	2	103, 97
rps2	4	101, 109, 99, 390
rps3	2	409, 102
rps8	3	327, 95, 277
tufA	3	95, 100, 110

The introns in the size range 95-107 nt are examples of the newly described group III introns (Christopher and Hallick, 1989). Some of the introns fall outside of the size range for group III, and have secondary structure features incompatible with group II. For example, the 409-nt *rps3* intron 1 has 5'- and 3'-boundary sequences more characteristic of group III than of group II, and the location of domains V and VI appear to lie 20 nt upstream from the beginning of *rps3* exon 2. We have postulated that *rps3* intron 1 might be a 99-nt group III intron with an internal 300-nt group II intron (Christopher and Hallick, 1989). There is a precedent for an intron-within-an-intron, which we have designated as "twintron". The 1042-nt intron of the *Euglena* chloroplast *psbF* locus has been shown to be a group II intron within another group II intron (Copertino and Hallick, 1991). Another example of a novel intron is the 208-nt *rpl16* intron 3, which is exactly twice the expected size of a group III intron, and lacks any group II-like secondary structure. It might represent one group III intron within another group III intron. We are currently characterizing the RNA maturation pathways for several chloroplast ribosomal protein genes to test this twintron hypothesis.

ACKNOWLEDGEMENTS

This work was supported by grants from the U. S. Public Health Service, National Institutes of Health.

REFERENCES

Copertino, D. W., and R. B. Hallick (1991). "Group II Twintron: An intron-within-an-intron in a chloroplast cytochrome b-559 gene", EMBO J. 10(2), in press.

Christopher, D. A., J. C. Cushman, C. A. Price and R. B. Hallick (1988). "Organization of ribosomal protein genes *rpl23*, *rpl2*, *rps19*, *rpl22* and *rps3* on the *Euglena gracilis* chloroplast genome." Curr Genet. 14(3): 275-85.

Christopher, D. A. and R. B. Hallick (1989). "*Euglena gracilis* chloroplast ribosomal protein operon: a new chloroplast gene for ribosomal protein L5 and description of a novel organelle intron category designated group III." Nucleic Acids Res. 17(19): 7591-608.

Christopher, D. A. and R. B. Hallick (1990). "Complex RNA maturation pathway for a chloroplast ribosomal protein operon with an internal tRNA cistron." The Plant Cell 2, 659-671.

Hudson, G. S., Mason, J. G., Holton, T. A., Koller, B., Cox, G. B., Whitfeld, P.R., and Bottomley, W. (1987). "A gene cluster in the spinach and pea chloroplast genomes encoding one CF_1 and three CF_0 Subunits of the H^+-ATP synthase complex and the ribosomal protein S2", J. Mol. Biol. 196: 283-298.

Manzara, T. and R. B. Hallick (1987). "Nucleotide sequence of the *Euglena gracilis* chloroplast gene for ribosomal protein L20." Nucleic Acids Res. 15(9): 3927.

Michel, F., K. Umesono and H. Ozeki (1989). "Comparative and functional anatomy of group II catalytic introns--a review." Gene. 82(1): 5-30, 1989.

Montandon, P. E., A. C. Knuchel and E. Stutz (1987). "*Euglena gracilis* chloroplast DNA: the untranslated leader of *tufA*-ORF206 gene contains an intron." Nucleic Acids Res. 15(19): 7809-22.

Montandon, P. E. and E. Stutz (1983). "Nucleotide sequence of a *Euglena gracilis* chloroplast genome region coding for the elongation factor Tu; evidence for a spliced mRNA." Nucleic Acids Res. 11(17): 5877-92.

Montandon, P. E. and E. Stutz (1984). "The genes for the ribosomal proteins S12 and S7 are clustered with the gene for the EF-Tu protein on the chloroplast genome of *Euglena gracilis*." Nucleic Acids Res. 12(6): 2851-9.

Nickoloff, J. A., D. A. Christopher, R. G. Drager and R. B. Hallick (1989). "Nucleotide sequence of the *Euglena gracilis* chloroplast genes for isoleucine, phenylalanine and cysteine transfer RNAs and ribosomal protein S14." Nucleic Acids Res. 17(12): 4882.

Orozco, E. M. Jr. and R. B. Hallick (1982). "*Euglena gracilis* Chloroplast Transfer RNA Transcription Units". II. Nucleotide Sequence Analysis of a tRNA-Val-tRNA-Asn-tRNA-Arg-tRNA-Leu Gene Cluster", J. Biol. Chem. 257, 3265-3275.

Passavant, C. W. , G. L. Stiegler and R. B. Hallick (1983). "Location of the Single Gene for Elongation Factor Tu on the *Euglena gracilis* Chloroplast Chromosome", J. Biol. Chem. 258, 693-695.

RPS10 AND 6 OTHER RIBOSOMAL PROTEIN GENES FROM THE S10/*SPC*-OPERON NOT ENCOUNTERED ON HIGHER PLANT PLASTID DNA ARE LOCATED ON THE CYANELLE GENOME OF *CYANOPHORA PARADOXA*

W. Löffelhardt, [1]C.Michalowski, M. Kraus, B. Pfanzagl, C. Neumann-Spallart, J. Jakowitsch, M. Brandtner and [1]H.J. Bohnert
Institut für Allgemeine Biochemie und
Ludwig-Boltzmann- Forschungsstelle für Biochemie
Universität Wien
Währingerstraße 38
A-1090 Wien
Austria

ORGANIZATION OF CYANELLE RIBOSOMAL PROTEIN GENES

The cyanelles from the flagellated protist *Cyanophora paradoxa* constitute a plastid type *sui generis*. Cyanelles are surrounded by a murein layer and harbor a number of unusual genes in their 127 kbp genome (Wasmann *et al.*, 1987). In this report a 12 kbp region will be described that contains a number of clustered genes encoding ribosomal proteins the complete sequence of which has been determined recently (Evrard *et al.*, 1990a,b,c; Bryant and Stirewalt, 1990; Kraus *et al.*, 1990; Michalowski *et al.*, 1990; Bryant *et al.*, 1990). Figure 1 depicts the central part of the large single copy region of cyanelle DNA from *C. paradoxa 555 UTEX* showing genes for 18 ribosomal proteins, one translation factor, ferredoxin I and two tRNAs. In addition two ORFs are present, the significance of which has yet to be determined. These genes appear to be organized in an operon structure (*str* operon), a transcription unit comprising *rps18* and *rpl33* and a large gene cluster in which many genes from the prokaryotic S10 and *spc* operons are included. A comparison of the organization of these genes with those of cyanobacteria as the close relatives of cyanelles is possible to a limited degree only. The substantial progress in cyanobacterial molecular genetics during the past several years has been focussed mostly on the characterization of photosynthetic genes. Only for the *str* operon data are available. Cyanelle as well as algal plastid genomes contain *tufA*, encoding elongation factor Tu, which is in contrast to higher plant chloroplasts. The

[1]Department of Biochemistry, University of Arizona, Tucson, AZ 85721, USA

NATO ASI Series, Vol. H 55
The Translational Apparatus of Photosynthetic Organelles
Edited by R. Mache et al.

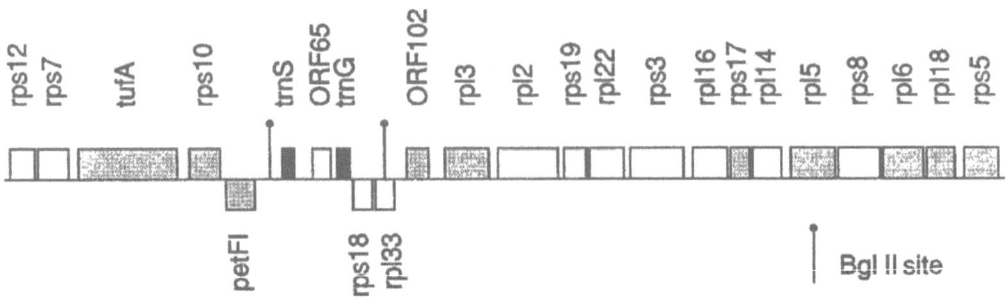

Figure 1: Ribosomal protein genes on a 12kbp domain of the cyanelle genome. Shaded areas indicate gene loci not present on higher plant plastid genomes. Genes represented by boxes above the line are transcribed from left to right, those below the line are transcribed from right to left.

gene specifying elongation factor G (*fus*) is absent from cyanelles and all plastids investigated. Nevertheless, due to the absence of introns in all cyanelle protein genes characterized so far, the cyanelle *str* operon resembles most closely that from *Anacystis nidulans* or *E. coli*. Identical gene content is also found in the plastid *str* operon of *Euglena gracilis* (Kraus *et al.*, 1990). An unusual feature, however is the recent identification of *rps10* very close to the 3' end of *tufA* (Neumann-Spallart *et al.*, unpublished; Bryant *et al.*, 1990). *rps10* has not yet been found in *A. nidulans*, whereas in *E. coli* it is located in a different operon. In higher plants and eukaryotic algae as far as the information is available, *rps10* is a nuclear gene . Surprisingly an analogous position of *rps10* adjacent to *tufA* and distal from the remainder of the "S10" operon has been reported for the archaebacterium *Methanococcus vannielii* (Lechner *et al.*, 1989; Auer *et al.*, 1989) where it is included in the transcript of the *str* operon . Thus far, we have not been able to obtain unequivocal proof for transcription of *rps10* in cyanelles, rendering it possible that *rps10* is a pseudogene. This would be the first observation of a non-functional gene in the densely packed cyanelle genome.

The *str* and the S10/*spc* operons are separated by about 15 kbp in *E. coli*. The thermophilic eubacterium *Thermotoga maritima* shows a very short inter-operon distance with *rps10* adjacent to *tufA* as in cyanelles and *M. vannielii* but contiguous with the S10 operon (O. Tiboni, personal communication). Figure 2 summarizes the different forms of organization of the *str* operons. Downstream of *rps10* with an intergenic distance of 72bp *petFI*, encoding cyanelle ferredoxin I, has been localized on the opposite strand (Neumann-Spallart *et al.*, 1990; Bryant *et al.*, 1990). This unique gene is followed by two tRNA genes separated by an ORF65 (specifying a 6.5kd hydrophobic protein of unknown function), that is also found on higher plant chloroplast genomes (Evrard *et al.*, 1990c). The adjacent genes appear to be in a different orientation again: *rps18* and *rpl33* (Evrard *et al.*, 1990a), suggesting a closely spaced transcription unit as it is known from plastid DNAs, whereas the respective genes in *E. coli* are unlinked. Downstream of *rpl33* separated by 123bp and on the other strand a functionally unidentified ORF102 is located in front of the large ribosomal protein gene cluster typical for plastid DNAs. In deviation from the plastid pattern the cyanelle cluster starts with *rpl3*, which is a nuclear gene in higher plants and green algae (Evrard *et al.*, 1990b). The subsequent 12 genes show short intergenic regions from 3bp (*rpl6-rpl18*) to 107bp (*rpl14-rpl5*) and follow the order observed in the large plastid cluster that is derived from the prokaryotic S10/*spc*/α operons (Evrard *et al.*, 1990a; Bryant and Stirewalt, 1990; Michalowski *et al.*, 1990).

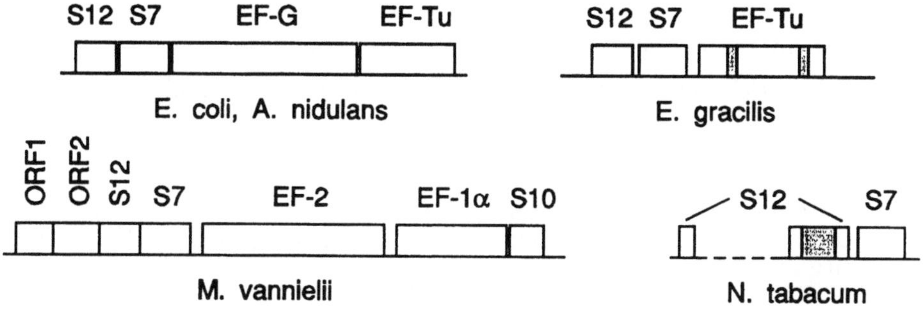

Figure 2: Organization of *str* operons from prokaryotes and plastids. Shaded regions mark introns. Elongation factor nomenclature with *M. vannielii* reflects the closer relationship to the eukaryotic counterparts (Matheson *et al.*, 1990).

In addition to *rpl3* five novel ribosomal protein genes, not encountered on the plastid genomes from higher plants (Sugiura, 1989) are present: *rps17*, *rpl5*, *rpl6*, *rpl18* and *rps5*. Missing from the cyanelle cluster is *rpl23*, the first gene in the plastid transcription unit, which has been shown to be a pseudogene in spinach (Thomas *et al.*, 1988). Also missing are the last four genes of the plastid cluster, namely *infA*, *rpl36*, *rps11* and *rpoA*. The separation of cyanelle *rps10* from "its" operon is peculiar, although it might be explained by a genome rearrangement by which a 2.7kb DNA segment, including *petFI*, *trnS*, ORF65, *trnG*, *rps18*, *rpl33* and ORF102 has been inserted between *rps10* and *rpl3*. One of these additional ribosomal protein genes, *rpl5*, is also found in the large gene cluster of the plastids from *E. gracilis* (Christopher and Hallick, 1989).

ANALYSIS OF INDIVIDUAL GENES

Initial data indicating the presence of ribosomal protein genes came from heterologous hybridization experiments. Gene probes for *rps12*, *rps7*, *tufA*, *rpl16*, *rpl14* and *rps8* from algae and higher plants were utilized. Thus the putative *str* operon and the large cluster were localized on BglII fragments 11 (5.1 kbp) and 5 (10.3 kbp), respectively (Löffelhardt *et al.*, 1990). These loci and their flanking regions were sequenced (Kraus *et al.*, 1990; Michalowski *et al.*, 1990). Genes *rps18*, *rpl33* and the 5' region of the large cluster were detected by continued sequencing downstream of *trnG* (Evrard *et al.*, 1990a). The comparison of open reading frames with sequences from data banks led to the identification of several genes which are arranged as shown in the figure 1. In general, all sequences encoding ribosomal proteins and translation factors were found to be well conserved, especially in the case of *rps12*, *tufA* and *rps19*. A pairwise comparison of coding regions for the subsequent genes from *rps12* to *rpl33* is shown in table 1.

Table 1: Percent amino acid identity of the cyanelle *str* and *rps18-rpl33* operon genes with the corresponding genes from selected organisms.

	rps12	rps7	tufA	rps10	rps18	rpl33
A. nidulans	91	56	82	-	-	-
E. gracilis	77	48	78	-	-	-
C. reinhardtii	85	-	75	-	-	-
N. tabacum	84	51	-	-	55	62
E. coli	76	44	71	51	41	32

Data are from Kraus *et al.*, 1990; Evrard *et al.*, 1990a.
(-): sequences not available

The highest identity scores were obtained with the cyanobacterium *A. nidulans*. At present no information on genes from the S10/*spc* operons is available for cyanobacteria. The cyanelle genes listed in table 2 are best matching their counterparts from the liverwort *M. polymorpha*. In addition we note that, with the exception of *rps17*, cyanelle genes appear to be more closely related to those from *Bacillus subtilis* than to those from *E. coli*. In higher plants sequence conservation is also considerable for nuclear genes: *tufA* from *Arabidopsis thaliana* is 74% identical to the cyanelle gene (Kraus *et al.*, 1990), and a pea *rps17* cDNA probe (provided by Dr. S. Gantt, Univ. of Minnesota) hybridized to the correct locus on cyanelle DNA. Very recently ORF105 located between *tufA* and *petFI* could be identified as *rps10* (figure 3).

```
                         str Operon | tuf A | TAATTAAAATCTA

AATTTTAGAGAGAATAGAAAATGTATTTATTTCTCTCTTTTTAATTTAAAAAATAAAAA

TAAAAATTAAAAATT ATG GCC AGT AAT CAA CAA TTA AAA ATT CGT ATT
                Met Ala Ser Asn Gln Gln Leu Lys Ile Arg Ile

CAA CTA CGT TCT TAT GAC TCT TCA TTA TTA GAA AAC TCA TGT GAA
Gln Leu Arg Ser Tyr Asp Ser Ser Leu Leu Glu Asn Ser Cys Glu

CAA ATT ATT GAA GCT GCA AAA CGA ACT GAT GCT ACA GCA GTA GGA
Gln Ile Ile Glu Ala Ala Lys Arg Thr Asp Ala Thr Ala Val Gly

CCT ATT CCT TTA CCA ACA AAA AAG AAA ATT TAT TGC GTT TTA CGT
Pro Ile Pro Leu Pro Thr Lys Lys Lys Ile Tyr Cys Val Leu Arg

TCC CCA CAT GTA GAT AAA GAT TCA CGA GAA CAT TTC GAA ATT AGA
Ser Pro His Val Asp Lys Asp Ser Arg Glu His Phe Glu Ile Arg

GTT CAT CGC AGA ATA ATT GAT TTA TAT TTA CCA TCT TCC AAA ACA
Val His Arg Arg Ile Ile Asp Leu Tyr Leu Pro Ser Ser Lys Thr

ATT GAT ACT TTA ACT CGT TTA GAT TTA CCA GCA GGT GTT GAT GTG
Ile Asp Thr Leu Thr Arg Leu Asp Leu Pro Ala Gly Val Asp Val

GAA GTT AAA CTT TAGTTTTTTTTGAATTAAG--inverted repeat (48bp)
Glu Val Lys Leu stop
```

Figure 3: Nucleotide sequence of cyanelle *rps10*.

An amino acid sequence comparison including S10 from the archaebacterium *Sulfolobus acidocaldarius* (29% identity with *C. paradoxa*, 27% identity with *E. coli*) is given in figure 4. Some highly conserved domains become apparent that also hold upon a comparison of the cyanelle

Table 2: Identity scores of cyanelle ribosomal protein genes of the "S10"- and *spc*-operon with homologs from other organisms

	rpl3	rpl2	rps19	rpl22	rps3	rpl16	rps17	rpl14	rpl15	rps8	rpl6	rpl18	rps5
eco	44	51	70	44	49	54	39	54	51	43	39	48	40
bsu/st	-	-	-	-	-	-	34	59	60	-	51	60	47
eug	-	63	55	30	24	63	-	66	55	27	-	-	-
mpo	-	65	72	53	48	73	-	75	-	57	-	-	-
nta	-	61	68	48	43	70	-	69	-	45	-	-	-

Data from Evrard *et al.*, 1990a,b; Michalowski *et al.*, 1990.
(-): sequences not available
eco - *E. coli*; bsu/st - *B. subtilis/stearothermophilus*; eug - *E. gracilis*; nta - *N. tabacum*; mpo - *M. polymorpha*

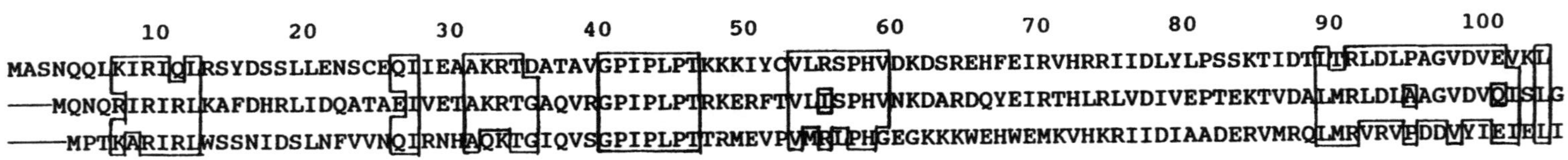

Figure 4: Amino acid sequence comparison of ribosomal protein S10 from *C. paradoxa* (upper line, *E. coli* (mid line) and *S. acidocaldarius* (lower line; data from Matheson *et al.*, 1990)

protein with S10 from *Mycoplasma capricolum* (Bryant *et al.*, 1990). In one case gene identification could be corroborated by detection of the corresponding gene product. Heterologous antibodies directed against ribosomal protein L3 from *E. coli* reacted with a protein band on Western blots obtained from cyanelle ribosome preparations. The same protein band was stained when *E. coli* extracts after transformation with cyanelle *rpl3* were blotted and reacted with this antibody (Evrard *et al.*, 1990b).

TRANSCRIPT ANALYSIS

The genes outlined in figure 1 give rise to 3 transcripts, that are processed to different extent: the transcript of the *str* operon including genes from *rps12* to (possibly) *rps10*, a bicistronic mRNA for *rps18* and *rpl33* and a large transcript comprising all genes from *rpl3* to *rps5* (table 3).

Table 3: Results from Northern hybridization of selected gene probes to cyanelle RNA. The size of minor signals is given in parentheses.

gene probe used	transcript size (kb)	transcription unit
rps12 (+rps7)	1.5 (3.4,4.2)	str operon
tufA	1.45 (3.4,4.2)	str operon
rps10	no signal	? str operon
petFI	0.45	monocistronic
ORF65	0.43	monocistronic
rps18 (+rpl33)	0.7 (1.1)	rps18-rpl33
ORF102	no signal	
rpl3	7.5	large cluster
rpl2	7.5 (3.0,2.4)	large cluster
rps19	7.5 (6.5,5.0,3.0,2.4)	large cluster
rpl16	7.5 (6.5,5.0,3.0)	large cluster
rps3	7.5 (6.5,5.0,3.0)	large cluster
rpl18	7.5 (3.0,2.4,1.8)	large cluster

Two major transcripts have been observed for the *str* operon. The primary transcript initiated from a promoter with only one deviation from the *E. coli* consensus sequences for the "-35" and "-10" boxes, is processed to a bicistronic mRNA including *rps12* and *rps7*, and to a *tufA* message, respectively (Kraus *et al.*, 1990). There is no promoter-like sequence between *tufA* and *rps10*. We thus assume, although the experimental proof is still missing, that *rps10* is cotranscribed with the *str* operon genes. A perfect inverted repeat with a 21bp stem and a 4b loop situated between *rps10* and *petFI* might function as a transcription termination signal (Neumann-Spallart *et al.*, 1990). The adjacent genes, encoding proteins which are not (*petFI*) or most likely not (ORF65) involved in functions of the translation apparatus, give rise to monocistronic transcripts (Neumann-Spallart *et al.*, 1990; Evrard *et al.*, 1990c) No signal on RNA blots could be obtained for ORF102 (Evrard *et al.*, 1990b). S1 mapping indicated a transcription start for the genes of the large ribosomal protein gene cluster at approximately 90 bp upstream of the *rpl3* start codon (Evrard *et al.*, 1990b). There is a good Pribnow box at the appropriate distance but no well defined "-35" box. Sequences indicating promoters internal to the S10/*spc* cluster could not be found. Transcription of this cluster may be terminated by an inverted repeat downstream of *rps5* (Michalowski *et al.*, 1990). The adjacent gene, ORF492, does not encode a ribosomal protein and is transcribed independently (Michalowski *et al.*, in preparation). Upon Northern hybridization six different gene probes from the large cluster all yielded a significant signal at approximately 7.5kb and minor, defined bands in variable sizes and intensities. This feature indicates a pattern for posttranscriptional processing which is intermediate between spinach chloroplasts, where only long, stable transcripts have been observed (Zhou *et al.*, 1988), and *E. gracilis* plastids, where the primary transcript is rapidly converted into smaller, gene-size mRNAs (Christopher and Hallick, 1989). Putative ribosome binding sites could be ascribed to all ribosomal protein genes with the exception of *rps7*, *rps10* and *rps19*. This might explain the stability of the long primary transcript, as individual genes may be recognized by the ribosome for the initiation of protein synthesis, whereas in *E. gracilis* ribosome binding sites are not easily recognized, or are absent.

THE RIBOSOMAL GENE COMPLEMENT OF THE CYANELLE GENOME

At present sequence information is available for about 40% of the cyanelle genome (D. Bryant, personal communication). Nevertheless one may infer that the number of cyanelle ribosomal protein genes exceeds that found in completely sequenced chloroplasts. *N. tabacum* and *M. polymorpha* have 20 genes in common plus one unique gene for each plant. This gives a total of 22 plastid ribosomal protein genes. A DNA region corresponding to less than one tenth of the cyanelle genome already contains 18 such genes seven of them being novel. The sequences of two other cyanelle ribosomal protein genes, *rpl20* and *rpl35*, have recently been published (Bryant and Stirewalt, 1990). Genes *rpl36* and *rps11* while missing from the cyanelle large cluster have been mapped by heterologous hybridization on different regions of the cyanelle genome (M. Götz, unpublished). A pea cDNA probe for *rpl24* (received from Dr. S. Gantt, Univ. Minnesota) gave a signal with cyanelle DNA but not with nuclear DNA from *C. paradoxa*. Sequence information indicates the presence of some more cyanelle ribosomal protein genes (D. Bryant, personal communication). On the other hand there is only one case, namely *rps14*, where a documented plastid gene could not be assigned to the cyanelle genome (M. Götz, unpublished). In summary the study of ribosomal protein gene organization has revealed more peculiarities of the cyanelle genome than have been found with respect to genes of the photosynthetic apparatus. Doubtless the unique plastids of *C. paradoxa* will continue to serve as an interesting model system for the study of organelle evolution.

Acknowledgements: This work was supported by grants from the Fonds zur Förderung der wissenschaftlichen Forschung (S29/06, WL), the Arizona Agricultural Experimental Station (#17441, HJB), and from NSF-International Programs. We thank Barbara Hamilton for help with the preparation of the manuscript. We are indebted to Don A. Bryant (Penn. State Univ.) and to Marcel Kuntz (IBMP, Strasbourg) for communicating results on cyanelle gene organization prior to publication. The *rps10* sequence has been deposited in the EMBL, GenBank and DDBJ databases under the accession number X52143.

REFERENCES

Auer J, Lechner K, Böck A (1989) Gene organization and structure of two transcriptional units from *Methanococcus* coding for ribosomal proteins and elongation factors. Can J Microbiol 35:200-204

Bryant DA, Stirewalt VL (1990) The cyanelle genome of *Cyanophora paradoxa* encodes ribosomal proteins not encoded by the chloroplast genomes of higher plants. FEBS Lett 259:273-280

Bryant DA, Schluchter WM, Stirewalt VL (1990) Ferredoxin and ribosomal protein S10 are encoded on the cyanelle genome of *Cyanophora paradoxa*. Gene, in press

Christopher DA, Hallick RB (1989) *Euglena gracilis* chloroplast ribosomal protein operon: a new chloroplast gene for ribosomal protein L5 and description of a novel organelle intron category designated group III. Nucleic Acids Res 17:7591-7608

Evrard JL, Kuntz M, Weil JH (1990a) The nucleotide sequence of five ribosomal protein genes from the cyanelles of *Cyanophora paradoxa*: implications concerning the phylogenetic relationship between cyanelles and chloroplasts. J Mol Evol 30:16-25

Evrard JL, Johnson C, Janssen I, Löffelhardt W, Weil JH, Kuntz M (1990b) The cyanelle genome from *Cyanophora paradoxa*, unlike the chloroplast genome, codes for the ribosomal L3 protein. Nucleic Acids Res 18:1115-1119

Evrard JL, Weil JH, Kuntz M (1990c) An ORF potentially encoding a 6.5kd hydrophobic protein in chloroplasts is also present in the cyanellar genome of *Cyanophora paradoxa*. Plant Mol Biol, in press

Kraus M, Götz M, Löffelhardt W (1990) The cyanelle *str* operon from *Cyanophora paradoxa*: sequence analysis and phylogenetic implications. Plant Mol Biol, in press

Lechner K, Heller G, Böck A (1989) Organization and nucleotide sequence of a transcriptional unit of *Methanococcus vannielii* comprising genes for protein synthesis elongation factors and ribosomal proteins. J Mol Evol 29:20-27

Löffelhardt W, Kraus M, Pfanzagl B, Götz M, Brandtner M, Markmann-Mulisch U, Subramanian AR, Michalowski C, Bohnert HJ (1990) Cyanelle genes for components of the translation apparatus. In: Nardon P, Gianninazzi-Pearson V, Grenier AM, Margulis L, Smith DC (eds) Endocytobiology IV. INRA Paris, pp 561-564

Matheson AT, Auer J, Ramirez C, Böck A (1990) Structure and evolution of archaebacterial ribosomal proteins. In: Hill WE, Dahlberg A, Garrett RA, Moore PB, Schlessinger D, Warner JR (eds) The ribosome: structure, function, & Evolution. American Society for Microbiology Washington, pp 617-633

Michalowski CB, Pfanzagl B, Löffelhardt W, Bohnert HJ (1990) The cyanelle S10/*spc* operon from *Cyanophora paradoxa*. Mol Gen Genet, in press

Neumann-Spallart C, Brandtner M, Kraus M, Jakowitsch J, Bayer MG, Maier TL, Schenk HEA, Löffelhardt W (1990) The *petFI* gene encoding ferredoxin I is located close to the *str* operon on the cyanelle genome of *Cyanophora paradoxa*. FEBS Lett 268:55-58

Sugiura M (1989) The chloroplast chromosomes in land plants. Annu Rev Cell Biol 5:51-70

Thomas F, Massenet O, Dorne AM, Briat JF, Mache R (1988) Expression of the *rpl23*, *rpl2* and *rps19* genes in spinach chloroplasts. Nucleic Acids Res 16:2461-2472

Wasmann CC, Löffelhardt W, Bohnert HJ (1987) Cyanelles: organization and molecular biology. In: Fay P, Van Baalen C (eds) The Cyanobacteria. Elsevier Amsterdam New York Oxford, pp 303-318

Zhou DX, Quigley F, Massenet O, Mache R (1989) Cotranscription of the S10- and *spc*-like operons in spinach chloroplasts and identification of three of their gene products. Mol Gen Genet 216:439-445

INTERACTION OF STREPTOMYCIN WITH 16S rRNA OF CHLOROPLASTS AND E.COLI

Erhard Stutz and Christophe Bonny
Laboratoire de Biochimie végétale
Université de Neuchâtel
CH-2000 Neuchâtel
Switzerland

Introduction

Chloroplast ribosomes have many structural and functional features in common with bacterial type ribosomes (70S) notably the sensitivity towards a wide range of antibiotic substances which interfere with one or several steps of protein synthesis. It is known for a long time that e.g. streptomycin (Sm) specifically blocks protein synthesis within chloroplasts without affecting the function of cytosolic ribosomes. Prolonged exposure to Sm leads, e.g., to irreversibly bleached Euglena gracilis cells lacking photosynthetically active chloroplasts. On the other hand several Sm-resistant or Sm-dependent mutants were described both for algae (Nicolas,1981, Boschetti and Bogdanov,1973, Bartlett et al,1979) and higher plants (Maliga et al,1975) and it was argued that such mutations occurred on the chloroplast genome, however, without characterization of the mutated loci on the nucleotide level.

A rigorous identification of antibiotic resistance conferring mutations became possible with the elucidation of chloroplast genomes on a molecular level. Furthermore, work with bacterial systems, e.g. E. coli, revealed that not only certain mutated ribosomal proteins but also specific mutations of rRNA conferred antibiotic resistance. This implicates that rRNA is directly involved in almost every step of translation and not only in the initial rRNA recognition step (Shine-Dalgarno sequence). Newer protocols, like e.g. Sm cross-linking studies (Gravel et al, 1987) or nucleotide modification protection experiments using as chemical probes either dimethylsulfate or kethoxal (Moazed and Noller,1987) allowed to locate sites on 16S rRNA which interact with Sm. Accordingly, a highly conserved sequence around position 912-915 (E. coli) is the primary target site for Sm. A second site of weaker interaction involves bases in position

NATO ASI Series, Vol. H 55
The Translational Apparatus of Photosynthetic Organelles
Edited by R. Mache et al.

1413, 1487 and 1494. Indeed the two a priori distal domains undergo cross-linking with Sm what may indicate that the two regions are proximal in the functional 30S subunit.

Dominance of Sm-sensitivity over resistance

The functional importance of the central 912-915 region of 16S rRNA is documented through genetic data and sequencing studies. It was shown that at least two Sm-R mutants of Euglena carry a C to U transition at position 876 which is equivalent to position 912 of E.coli (Montandon et al,1985) and similar observations were made with Sm-R Chlamydomonas (Harris et al,1989) and tobacco (Fromm et al,1989) mutants. But surprisingly, no Sm-R E.coli mutants of this kind were found, though specific attempts were made (Morgan et al,1988). Breckenridge and Gorini (1969) already noticed that Sm-sensitivity is dominant over resistance in growing cultures, and the genetic make up (seven rrn operons) seems not to allow in vivo selection of Sm-R E.coli mutants with mutated 16S rRNA. Morgan et al. (1988) tried oligonucleotide directed mutagenesis without success. However, two reports appeared claiming the isolation of Sm-R E.coli mutants, one mutation being at position 912 (C to T) (Montandon et al,1986) the other at position 523 (A to C) (Melançon et al,1988). In both cases similar screening protocols were used introducing the 16S RNA mutations by transforming E.coli with multicopy plasmids carrying the relevant mutations. There is however a basic difference in the two kinds of Sm-R mutants, i.e. the mutation (C912-T912) prevents Sm to bind in this very region, while the "530" region is not a Sm-binding site and a mutation (A523-C523) seems not to prevent Sm binding on 16S rRNA, it rather counteracts its adverse function.

The selection protocol used in both cases does not preclude the possibility that additional mutations occur which may confer drug resistance. In order to test this possibility we reexamined our E.coli mutant 50.109 (C912-T912) for possible Sm-resistance conferring mutation(s) in the S12 gene. The strategy was to insert a 4.4 kbp EcoR1 fragment containing the S12 gene from mutant 50.109 into pBR322 and to use this construct (pS12) for sequencing and transformation experiments (Bonny et al, 1990, submitted). The sequencing experiment (not shown) revealed that a mutation had occurred at codon position 42 of the S12 protein (Lys to Thr) rendering thereby the ribosomes and therefore the genotype Sm-resistant (Wittmann and Wittmann-Liebold,1974).

It was of interest to test the transformation capacity of construct pS12 using both E.coli HB101 (Sm-R) and DH1 (Sm-S) and to compare it with other transformants as shown in **Table 1.** Two series of experiments were made: A) Transformed cells were plated on agar containing 100 µg/ml Sm and colonies scored after 24 h of incubation; B) Transformants were grown to log phase in the presence of ampiciline without Sm. Aliquots were spread on agarose plates having previously been incubated for 1 h in liquid medium with (100 µg Sm/ml) or without Sm (control, not shown). After 24 h of incubation, colonies were counted. We see that a) the construct pS12 does not render HB101 cells Sm-sensitive, nor does it convert DH1 to a Sm-resistant mutant (A) but it increases considerably the survival capacity of DH1 (B); b) the pKK-C914 transformant behaves identically, while the pBR322 transformed cells did not grow on Sm containing plates (A) nor survive a single short Sm exposure (B). Note, that none of these transformants had undergone a screening procedure, i.e. an adaptation phase. In contrast to these transformants stands the DH1 derived mutant 50.109 (Montandon et al,1986) which behaves like HB101.

Table 1. Capacity of E.coli transformants to grow in SM-containing medium (A) and to survive after a single Sm incubation (B)

E. coli	transformed with construct	A*	B**
HB101 (Sm-R)	pS12	yes	yes
DH1 (Sm-S)	pS12	no	yes
DH1	pBR322	no	no
DH1	pkk-C914	no	yes
DH1 (50.109)	pkk-T912	yes	yes

*) medium with 100 µg Sm/ml
*) exposure to 100 µgSm/ml for 1 h (see text)

We studied this behaviour on the ribosomal level using E.coli transformed with pKK3535 derived plasmids carrying, respectively, a C912-T912 (pT912), a A914-C914 (pC914) and a C912-T912:G888-A888 (pT912:A888) mutation. The latter mutation was constructed to test whether restoring the possibility of base pairing would also restore Sm-sensitivity. Positive transformants were subdivided in two classes : Class A are transformants which did not undergo further screening, i.e. series of replica platings in the presence of increasing concentrations of Sm as published (Montandon et al,1986). Class B are transformants which underwent replica-platings in the presence of Sm. From both classes of cells we isolated ribosomes which were characterized with respect to polymerization index (I) and misreading index (M). The results are given in **Table 2.** We see that ribosomes from class A mutants are about 3 times as

Table 2. Incorporation index (I) and misreading index (M) of ribosomes from E.coli

Strain	Plasmid	Class	I(%)	M1 a	M1 b	M1 a/b	M2 a	M2 b	M2 a/b
DH1	pKK3535	-	20	27	5	5.4	7	1	7.0
	pT912:A888	A	40	23	5	4.6	-	-	-
	pT912	A	55	19	5	3.8	-	-	-
	pC914	A	65	19	4	4.8	-	-	-
DH1	pT912:A888	B	97	-	-	-	12	11	1.1
	pT912	B	101	-	-	-	11	10	1.1
	pC914	B	97	34	15	2.2	19	16	1.2
HB101	-	-	100	67	38	1.8	-	-	-

M1 and M2 are two independent series of experiments. a and b are without and with Sm, respectively; I(%) = cpm ^{3}H-Phe (+Sm)/ cpm ^{3}H-Phe (-SM); M = cpm ^{3}H-Phe/ cpm ^{14}C-Ile with 25 µg/ml Sm

active in vitro as are ribosomes from the control (DH1-pKK3535), but they can hardly restrict misreading. On the other hand ribosomes from class B cells are efficient in overall polymerization activity (I) in the presence of Sm and are resistant towards misreading, indistinguishable from ribosomes of the control (HB101). This strongly suggests that they have also acquired a genomic Sm-R S12 mutation like the mutant 50.109. These results are in line with the observation at a cellular level and can explain the survival capacity of transformants like DH1-pS12 and DH1-pKK-C914. Obviously, Sm-sensitivity is dominant in growing cultures and it can only be overridden by a homozygous Sm-R conferring mutation ,e.g. in the S12 gene. Liu et al. (1989) also observed that transformation of Sm-sensitive E. coli (TG-1) with a construct carrying a Sm-R chloroplast S12 gene from Chlamydomonas was unsuccessful.

Streptomycin interacts with the central domain of 16S rRNA

For specialists in ribosome research it came as a surprise when we reported that two Sm-R mutants of Euglena gracilis carried a point mutation in position 876 (equivalent to E. coli C912) of the chloroplast 16S rRNA and that this mutations most likely conferred Sm-resistance since no mutation in the chloroplast S12 gene had been found. This region of 16S rRNA had not attracted any particular attention at that time since none of the many results published hinted towards this region to be involved in

Table 3. Sm-resistance conferring mutations in chloroplast 16S rRNA

Organism strain	sequence position	E.coli equivalent	base shift	References
E.gracilis				
smr 139.12/4	876	912	C to T	Montandon et al,1985
smr 139.20/2	876	912	C to T	idem
C.reinhardtii				
sr-u-sm3	14	13	T to G	Harris et al,1989
sr-u-2-60	474	523	A to C	idem
sr-u-2-23	856	912	C to T	idem
sr-u-sm5	858	914	A to C	idem
sr-u-sm 3a	859	915	A to G	idem
E.eugametos				
sr-2		523	A to C	Gauthier et al,1988
N.tabacum				
Sr1	860	912	C to A	Etzold et al,1987
92/strR7	860	912	C to T	Fromm et al,1989
92/strR6	472	525	C to T	idem

translation steps perturbed by Sm. In the mean time, however, several Sm-resistant chloroplast mutants were described with altered 16S rRNA in the central region but also in the so-called "530" loop as listed in **Table 3**. This indicates that appropriate mutations in the 912-915 region do

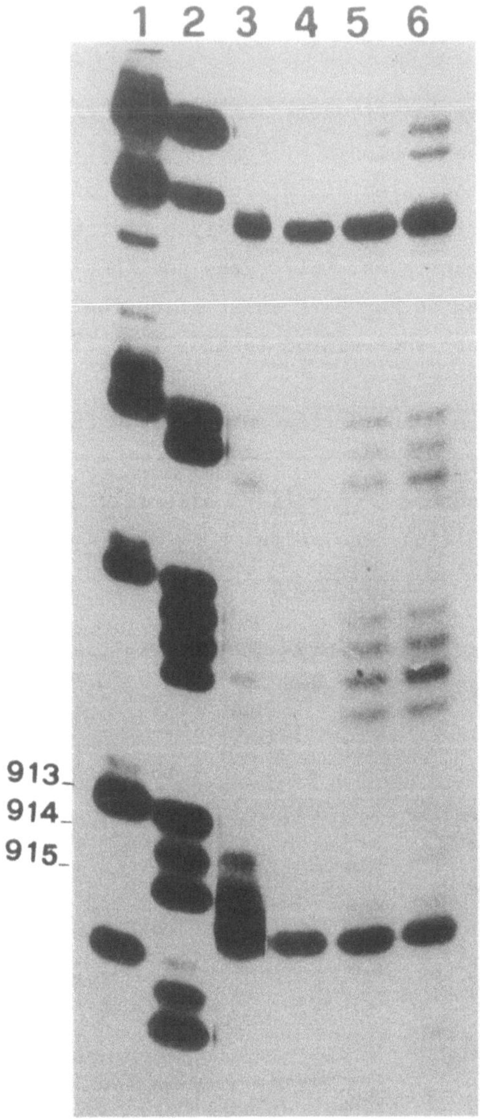

Fig.1. Chemical footprinting of the 912-915 region and its vicinity of 16S rRNA (E.coli) using dimethylsulfate (Moazed and Noller,1987).1) radiograph of unmodified wild-type rRNA; 2) A-sequence ladder; 3) modified 70S ribosomes from wild-type; 4) idem protected by Sm; 5) modified 70S ribosomes from mutant DH1-C914; 6) idem protected by Sm. Positions of the A triplet (lane 2) is given on the margin. Note that each band produced by a modified nucleotide is displaced by one nucleotide from the corresponding sequence band.

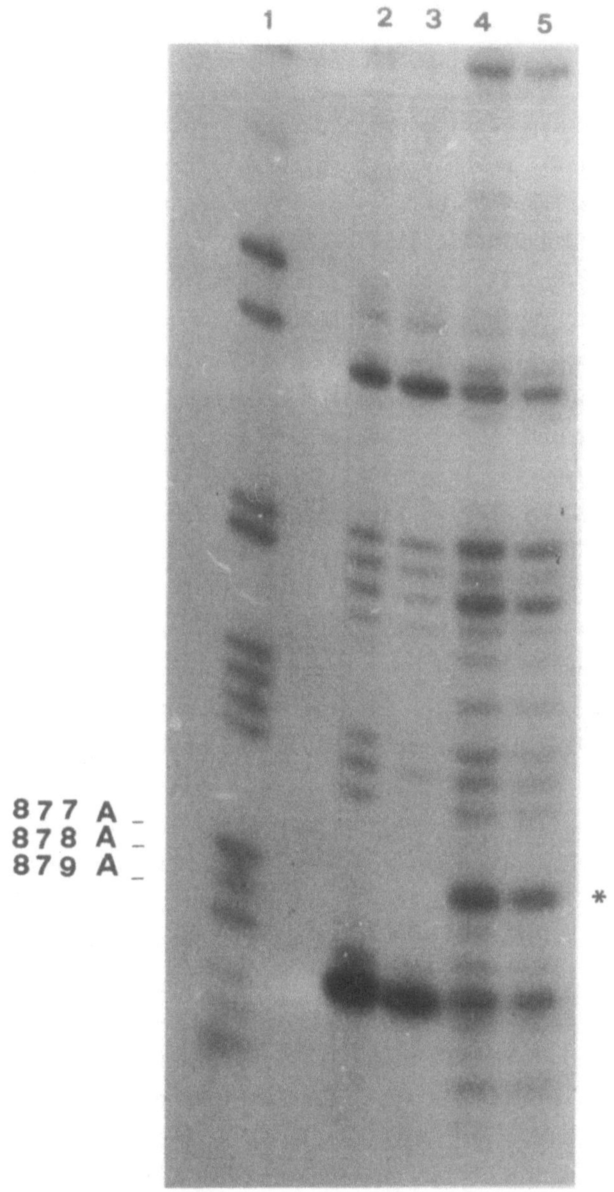

Fig.2. Chemical footprinting of the 876-879 region and its vicinity of Euglena gracilis chloroplast 16S rRNA with dimethylsulfate. 1) A-sequence ladder; 2) modified 70S ribosomes from wild-type; 3) idem protected by Sm; 4) modified ribosomes from mutant smr 139.12/4; 5) idem protected by Sm. *) marks unidentified band, see text.

confer Sm-resistance and stable Sm-R mutants can be obtained provided the genetic make up of the organisme (organelle) allows for a homozygous situation.

A most elegant approach to elucidate the role of particular 16S rRNA regions (nucleotides) in the diverse steps of translation was developed by Nollers group. They showed that, e.g. tRNAs (Moazed and Noller,1986) and antibiotics (Moazed and Noller,1987) could protect short regions or single bases from chemical modifications, e.g. interaction with dimethylsulfate or kethoxal. Protected regions are identified using a RNA sequencing protocol which identifies modified nucleotides. The chemical modification reactions are done with ribosomal subunits, intact ribosomes or rRNA.

We used this approach to characterize both E.coli and Euglena gracilis chloroplast ribosomes which carry wild type or mutated 16S rRNA (Bonny et al,1990, submitted). We show in **Fig. 1** a short segment of 16S rRNA which is known from Nollers experiments to interact with Sm. In line with their data we notice that Sm protects bases A913 to 915 (lane 4) and in addition we see that the A914-C914 mutation abolishes almost completely base methylation, independently of the presence of Sm (lanes 5, 6). This means that the mutation at position 914 not only prevents binding of Sm but renders the bases 913 to 915 essentially inaccessible to methylation.

We repeated the same experiment with chloroplast ribosomes from Euglena gracilis (**Fig. 2**). The equivalent positions which are identical in chloroplast 16S rRNA (adenines 877 to 879) also get protected by Sm, compare lane 2 with lane 3. Obviously, chloroplast ribosomes behave also in this respect similar to E.coli ribosomes. Furthermore we see that the C876-U876 mutation (lanes 4,5) also reduces the degree of methylation of bases A877, A878 and A879. We notice in lanes 4 and 5 the appearance of a strong band marked by an asterisk. This band does not correspond to A877. We have no explanation for the appearance of this band.

Mutations in two domains of 16S rRNA can confer Sm-resistance

Two domains of 16S rRNA were identified as crucial in the codon-anticodon recognition step, i.e., the "530"-loop and the 3' terminal region around position 1400. Switching G530 to any other base is lethal suggesting a direct catalytic function of this loop in translation (Powers and Noller,1990). As shown in **Table 3** Sm-resistance conferring mutations in chloroplast 16S rRNA "530" loop area were identified. Furthermore, Sm-R

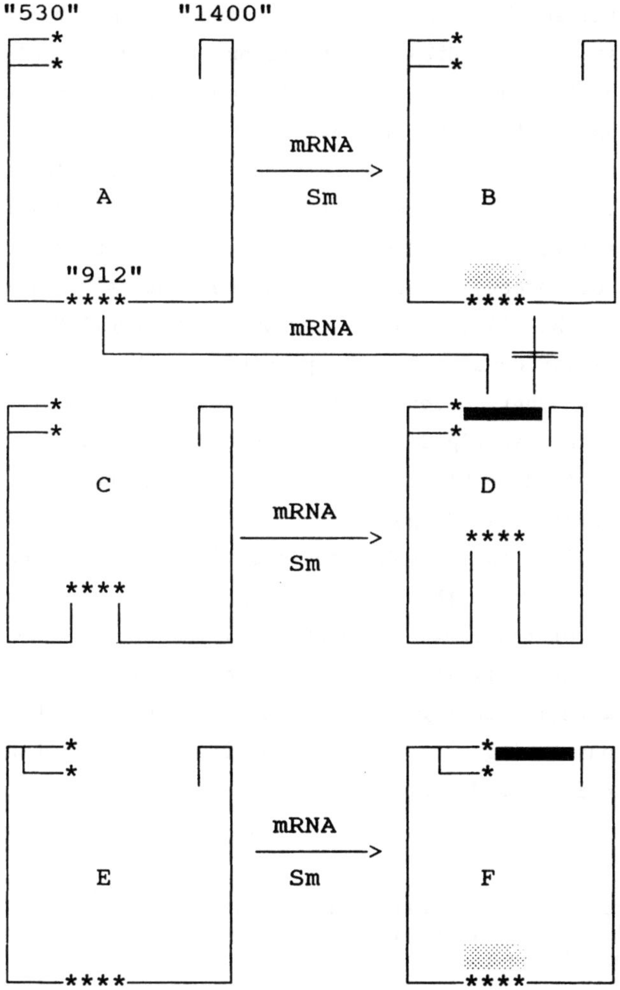

Fig.3. Graphic presentation of possible spatial shiftings of domains in 16S rRNA involved in Sm-resistance conferring mutations. A) inactivated conformer, wild-type; B) idem with bound Sm; C) inactivated conformer, Sm-resistant mutant (912-915); D) activated conformer due to transition either from A or C, B to C shift is blocked by Sm; E) inactivated conformer, Sm-resistant mutant ("530"-loop); F) activated conformer showing shift in the "530" domain compensating for the Sm-block in the "912-915" region. * signifies regions where Sm-resistance conferring mutations were identified; ▬ optimally spaced region involved in codon-anticodon recognition; ▒ bound streptomycin.

E.coli was obtained by transformation using a pKK3535 derived construct carrying a A523 to C523 transversion (Melançon et al,1988). No Sm-resistance conferring mutations in the 1400 region were reported yet. It seems that ribosomes from the Sm-R E.coli mutant (type C523) still bind Sm, contrary to ribosomes from Sm-R E.coli or chloroplast mutants (type 912-915 domain) which no longer interact with Sm. Such different behaviour of two domains towards Sm reflects a difference in function. We may assume that the read-out process requires a unique spatial conformation of the "530"-loop and the 1400-related region (Dahlberg,1989). The relative movement of the two domains is only possible if the central region (912 to 915) undergoes higher order spatial transitions during this step as graphically presentd in Fig.3 A to D. Sm blocks this pivotal function (B to D). Sm-R mutants (912 to 915) prevent binding of Sm but still allow for the required conformational change (C to D)to optimally position the catalytic domains. Quite differently Sm-R mutations in the "530"-loop may induce a conformational change in the very loop region which can compensate for the pivotal function of the central region, i.e. the Sm-block is neutralized (E to F).In that context it is interesting to note that both the "530" loop and the central region undergo pseudo-knotting (Gutell and Woese,1990) with mutated positions taking part in (523 and 525) or being adjacent to (912 to 915) the respective pseudo-knots.

Acknowledgement

We are grateful to C. Bettinelli for secretarial and S.Marc-Martin for technical help. This work is funded by Fonds National Suisse de la Recherche Scientifique (E.S.).

REFERENCES

Bartlett SG, Harris EH, Grabowy CT, Gillham NW, Boynton JE (1979) Ribosomal subunits affected by antibiotic resistance mutations at seven loci in Chlamydomonas reinhardtii. Mol Gen Genet 76:199-208

Bonny C, Montandon PE, Marc-Martin S, Stutz E (1990) submitted.

Boschetti A, Bogdanov S (1973) Different effects of streptomycin on the ribosomes from sensitive and resistant mutants of Chlamydomonas reinhardtii. Eur J Biochem 35:482-488.

Breckenridge L, Gorini L (1969) The dominance of streptomycin sensitivy re-examined. Proc Acad Sci USA 62: 979-985.

Dahlberg AE (1989) The functional role of rRNA in protein synthesis. Cell 57: 525-529.

Fromm H, Galun E, Edelman M (1989) A novel site for streptomycin resistance in the "530" loop of chloroplast 16S rRNA. Plant Mol Biol 12: 499-505.

Gutell RR, Woese CR(1990) Higher order structural elements in rRNAs: Pseudo-knots and the use of non-canonical pairs. Proc Natl Acad Sci USA 87: 663-667.

Gravel L, Melançon P, Brakier-Gingras (1987) Cross-linking of streptomycin to the 16S rRNA of E.coli. Biochemistry 26: 6227-6232.

Harris EH, Burkhart BD, Gillham NW, Boynton JE (1989) Antibiotic resistance mutations in the chloroplast 16S and 23S rRNA gens of Chlamydomonas reinhardtii. Genetics 123: 281-292.

Liu XQ, Gillham NW, Boynton JF (1989) Chloroplast ribosomal protein gene rps12 of Chlamydomonas reinhardtii. J Biol Chem 264: 16100-16108.

Maliga P, Sz-Breznovits A, Marton L, Joo F (1975) Non Mendelian streptomycin resistant tobacco mutant with altered chloroplasts and mitochondria. Nature 255: 401-402.

Melançon P, Lemieux C, Brakier-Gingras L (1988) A mutations in the "530" loop of E.coli rRNA causes resistance to streptomycin. Nucleic Acids Res 16: 9631-939.

Moazed D, Noller HF, (1986) Transfer RNA shields specific nucleotides in the 16S rRNA from attack by chemical probes. Cell 47:985-994.

Moazed D, Noller HF (1987) Interaction of antibiotics with functional sites in 16S rRNA. Nature: 389-394.

Montandon PE, Nicolas P, Schurmann P, Stutz E (1985) Streptomycin resistance of Euglena gracilis chloroplasts: identification of a point mutation in the 16S rRNA gene in an invariant positiόn. Nucleic Acids Res 13:4299-4310.

Montandon PE, Wagner R, Stutz E (1986) E. coli ribosomes with a C912 to U base change in the 16S rRNA are streptomycin resistant. EMBO J 5: 3705-3708.

Morgan EA, Gregory ST, Sigmund CD, Borden A (1988) Antibiotic resistance mutations in E.coli rRNA genes and their uses, in: NATO series (Tuite MF, Picard M, Bolotin-Fukahara M, eds) vol. H14, pp 43-53, Springer-Verlag, Berlin.

Nicolas P (1981) Sensitivity of Euglena gracilis to chloroplast-inhibiting antibiotics and properties of antibiotic-resistant mutants. Plant Sci Lett 22: 309-316.

Powers T, Noller HF (1990) Dominant lethal mutations in a conserved loop in 16S rRNA. Proc Natl Acad Sci USA 87:1042-1046.

Thomas CL,Gregory RJ, Winslow G,Muto A, Zimmermann RA (1988) Mutations within the decoding site of E.coli 16S rRNA: growth rate impairment, lethality and intragenic suppression. Nucleic Acids Res 16: 8129-8146.

Wittmann HG, Wittmann-Liebold B (1974) Chemical structure of bacterial ribosomal proteins. in: Ribosomes (Nomura M, Tissieres A, Lengyel P, eds) pp.115-140, Cold Spring Harbour, Monograph Series.

THE STRUCTURE OF THE ANTIBIOTIC BINDING SITES IN BACTERIAL RIBOSOMES

JUAN P.G. BALLESTA
Centro de Biologia Molecular,
CSIC and UAM,
CantoBlanco,
28049 Madrid,
Spain.

With only a few exceptions, antibiotics inhibit protein synthesis by binding to the ribosome and blocking some of the several functions that the particle performs in this metabolic process (Vazquez,1979; Gale et al. 1981) . The antibiotic binding sites must be located in regions of the ribosomal particle closely related to the inhibited activity and, obviously, the definition of the structure of these interaction sites will provide important information for the understanding of both the drug mode of action and the function of the ribosome.

In fact, the study of antibiotics in general cannot be dissociated from that of their targets, and this is especially true in the case of protein synthesis inhibitors, where drugs such as puromycin, chloramphenicol or streptomycin have been used as both classical and fundamental tools in the investigation of this synthetic pathway.

In this report an overview of the studies that are presently being carried out by several laboratories on the structure of antibiotic interaction sites in the ribosome will be presented, discussing in detail some representative cases which can illustrate some of the controversial issues of the field and the possible future developments.

NATO ASI Series, Vol. H 55
The Translational Apparatus of Photosynthetic Organelles
Edited by R. Mache et al.

METHODS OF ANALYSIS OF ANTIBIOTIC BINDING SITES

In order to identify the ribosomal components involved in the drug interation, the structure of the antibiotic sites has been studied using basically three main approaches: a, affinity labeling; b, chemical and enzymatic footprinting; and c, analysis of antibiotic resistant species. As we will comment later, each one of these techniques has advantages and disadvantages, but only by combining data from the different methodological approaches can a real understanding of the antibiotic interaction sites will be achieved.

Affinity labeling

This method (see Cooperman , 1980 for a detailed discussion), especially when photoactivated compounds are used, is the most direct approach since, in principle, the drug would react directly with the ribosomal components forming its actual interaction site. However, precautions have to be taken when interpreting the results from affinity labeling experiments since there are several drawbacks to this technique. Thus, only a minor fraction of the bound drug, 10% in the best conditions, reacts covalently with the receptor, and this low yield may raise questions about the significance of the detected labeling . In addition, the labeling at non-specific sites cannot be totally ruled out since activation of the molecules obviously not only takes place at the binding site and the activated molecules can also react with other ribosomal components. Nevertheless, if appropriate controls are used, these problems can be overcome and , in fact, results from affinity labeling experiments when properly performed, provide very relevant data for defining the structure of the antibiotic binding sites in the ribosome (Cooperman 1980).

Footprinting

The analysis of the alterations in the reactivity of the ribosomal components induced by ligands, generally called "footprinting" techniques, can give useful structural information on the interaction sites. In the case of antibiotics, it has been applied almost exclusively to the study of the rRNA . By determining the rRNA modification pattern when the ribosome is treated with certain chemical reagents and enzymes in the presence and in the absence of an antibiotic, it has been possible to relate the changes in the chemical reactivity of certain nucleotides to the interaction of the drug (Moazed and Noller,1987a, Egebjerg et al. 1989). It may be assumed that the nucleotides, showing a modified reactivity, must be closely related to the drug interaction site. However, considering the high cooperativity of the ribosomal structure, an allosteric effect of the drug interaction is a possibility that cannot be excluded.

Analysis of resistant mutants

Since resistance to the antibiotics is frequently due to an alteration of the target structure that reduces its affinity for the drug, the presence of a modified ribosomal component in resistant particles is usually interpreted as an indication of its involvement on the drug interaction site (Cundliffe,1989). The cooperativity of the ribosome structural components also argues against too straightforward an interpretation of the results of this approach. This precaution is especially important in this case, since in some instances, binding of the drug to the resistant ribosomes with an affinity similar to that the sensitive particles has been reported (Jimenez and Vazquez,1975), suggesting that the altered component might not be directly involved in the drug binding site .

The fragmnentary knowledge of the ribosomal structure is also a very important limitation for the correct interpretation of the results obtained from any of the above-mentioned techniques. Thus, with few exceptions, the ribosome structure models assume a globular conformation for ribosomal proteins, which in fact are represented as rigid spheres (Schüler and Brimacombe, 1988; Walleczek et al. 1988) . Although this is an openly assumed over-simplification of reality, it may lead us to consider as "unexpected" some results that with a more realistic view of the protein structure may be perfectly compatible with present day accepted models. Similar comments can apply to the conclusions based on our present limited knowledge of the rRNA tertiary structure .

ANALYSIS OF DATA ON THE STRUCTURE OF SOME ANTBIOTIC BINDING SITES

The number of ribosomal antibiotic binding sites studied in some detail is rather small, and only a few have been analyzed using more than one experimental approach. In some cases the agreement among the results from the different methods is good, but on other occasions there are discrepancies and it is difficult to say whether they are due to artefacts of the procedures or to the limitation of our knowledge of the ribosome structure .

Rather than make a general review of the field that would require a substantially more extensive report, some representative cases will be presented in some detail to illustrate the different situations encountered in the field and provide those interested with a general view of the more problematical issues.

Thiostrepton

Thiostrepton is probably one of the best examples of consistent results from different approaches . Data obtained by either chemical protection or analysis of resistant ribosomes have allowed a reasonable picture of the drug ribosomal binding site to be drawn.

Thus, the data indicating that the drug binding site was defined by a protein -RNA complex formed by protein L11 and the 23S rRNA fragment (Thompson et al. 1979) were afterwards confirmed showing that methylation of A_{1067}, located in the protein L11 binding domain, renders the ribosomes of *E. coli* thiostrepton-insensitive (Thompson et al. 1982) and a ribose-methylation at the equivalent position was responsible for drug insensitivity of the ribosomes in the thiostrepton producing organism *Streptomyces azureus* (Cundliffe and Thompson, 1979) . Moreover, mutagenesis of the nucleotide in either the same position *in Escherichia coli* (Thompson et al. 1988) or at the equivalent position in *Halobacterium halobium* (Hummel and Böck,1987) confers resistance to the drug. These results are in total correspondence with the nucleotide reactivity changes induced by the drug, that have been located exclusively within the nucleotides A_{1067}-A_{1098} , which define the L11 binding domain in the 23S rRNA (Egebjerg et al. 1989).

Thiostrepton binds, therefore, to the protein L11 interaction domain in the large subunit, and the presence of the protein,and at least one unmethylated adenosine at position 1067 are required for the right conformation of the drug binding site. The capacity of the free 23s rRNA to bind drug while the free protein L11 is unable to do so (Cundliffe, 1989) strongly suggests that thiostrepton interacts directly with the nucleic acid in the ribosome. Direct evidence of this fact is, however, lacking and, perhaps it could be provided by affinity labeling experiments which can be enhanced by the very high affinity ($K_A > 10^9$ M^{-1}) of thiostrepton for the ribosome. Unfortunately, reactive derivatives of this drug are not available at the moment.

The thiostrepton binding site is, on the other hand, in accordance with the antibiotic mode of action, since this drug has been reported to interfere with the function of the elongation factors (Vazquez, 1979: Gale et al. 1981), and the ribosomal interaction domain for these factors has been shown to overlap precisely with the protein L11 binding site (Girshovich et al. 1986).

We are still far from understanding at the molecular level the structure of the thiostrepton binding site, that obviously cannot be extended to the whole L11 binding domain nor be limited to the residue A1067. Nevertheless, the location of this site has clearly been narrowed down to a very small region of the ribosome structure and at least the major components have been identified.

Pactamycin

Pactamycin is a wide spectrum antibiotic active against prokaryotic and eukaryotic cells. Although it has little clinical interest due to its high toxicity , pactamycin is a very useful tool in protein synthesis research, since it is one of the very few drugs that is reported to inhibit protein synthesis initiation in eukaryotic systems (Vazquez, 1979; Gale et al. 1981).

The capacity of pactamycin to interact with ribosomes from bacteria and eukaryotic cells suggests that the drug interacts at a highly conserved site, and this fact confers an additional interest to the study of its structure.

The pactamycin binding site has , in fact, been analyzed using the three experimental approaches mentioned previously. In this case, however, the results do not totally agree.

Affinity labeling experiments performed using *E.coli* ribosomes indicated that in 70s bacterial ribosomes the drug interacts mainly with proteins S18 and S21 from the small subunit. In free 30S subunits protein S4 is labeled instead of S21 (Tejedor et al. 1985). The experiments were performed in conditions favoring the specificity of the labeling, but recently this point has been confirmed using ribosomes from sensitive and resistant *Streptomyces lividans* strains (Calcutt and Cundliffe, 1990). The experiments show that only in the first case does photoincorporation into the ribosomal proteins take place (Iglesias and Ballesta, unpublished results). It is therefore reasonable to assume that the labeled proteins are located at or near the pactamycin interaction site. Eukaryotic ribosomes have been also photolabeled with pactamycin ; protein S25 is the main labeled protein in rat liver (Synetos et al. 1986) and protein S21/24 and S10 in yeast (Tejedor et al. 1987). As discussed previously (Tejedor et al. 1985, 1987), these proteins can be related to the initiation step of protein synthesis in accordance with the postulated mode of action of the drug.

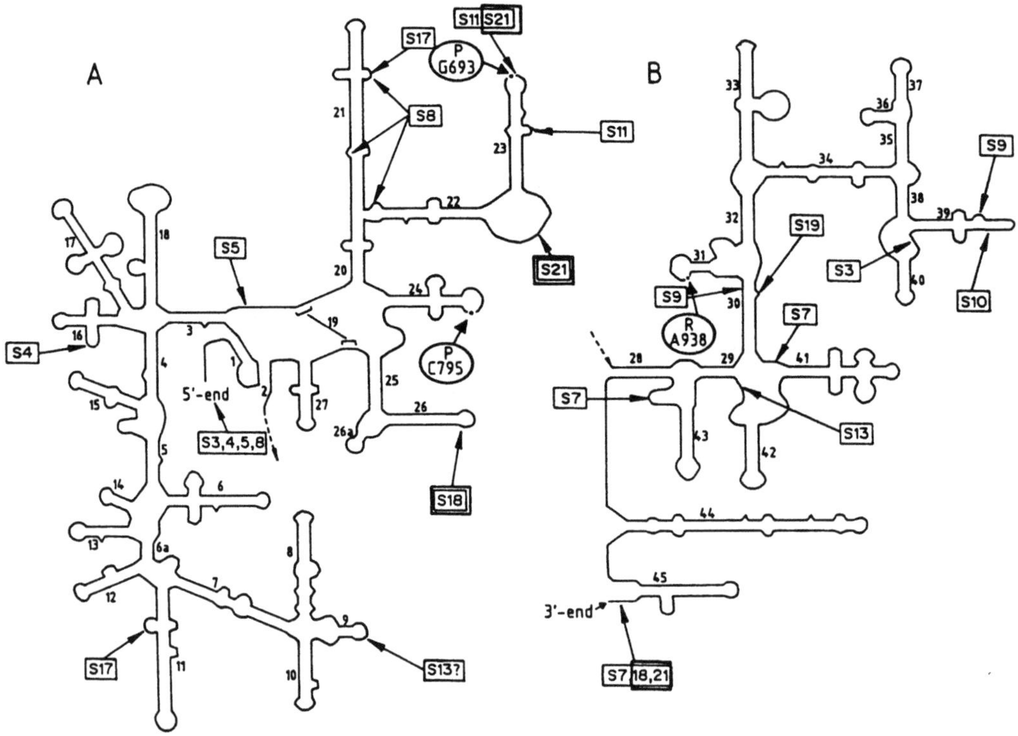

Figure 1.- Scheme of 16S rRNA secondary structure (Brimacombe et al. 1988). The interaction sites of a number of ribosomal proteins as well as the nucleotides protected by pactamycin (P) are marked. The position of the methylated residue in pactamycin resistant ribosomes (R) is indicated also.

Streptomyces pactum, the producer of pactamycin, has been shown to have resistant ribosomes, and it has been possible to clone a DNA insert from this organism, that confers pactamycin resistance to the sensisitve *Streptomyces lividans* (Calcutt and Cundliffe, 1990). The resistant strains contain a methylase activity that specifically introduces one methyl group into 16s 'rRNA. Similarly to what was done in other cases (Cundliffe, 1989), by means of RNAse H treatments and reverse transcriptase primer extension tests, we found that modification occurs at A_{938}, in the *Streptomyces* 16s rRNA, corresponding to A_{964} in the equivalent *E.coli* ribosomal component (Ballesta and Cundliffe, unpublished results). This nucleotide is located in a highly phylogenetically conserved region of the RNA (Raué et al.1988), which contains two additional modified nucleotides and that has been shown to be involved in subunit association (Herr et al. 1979) and also in the initiation step of protein synthesis (Tapprich et al. 1989). These two processes are affected by pactamycin and therefore these results are also compatible with the proposed mode of action of the drug (Vazquez, 1979 and Gale et al. 1981).

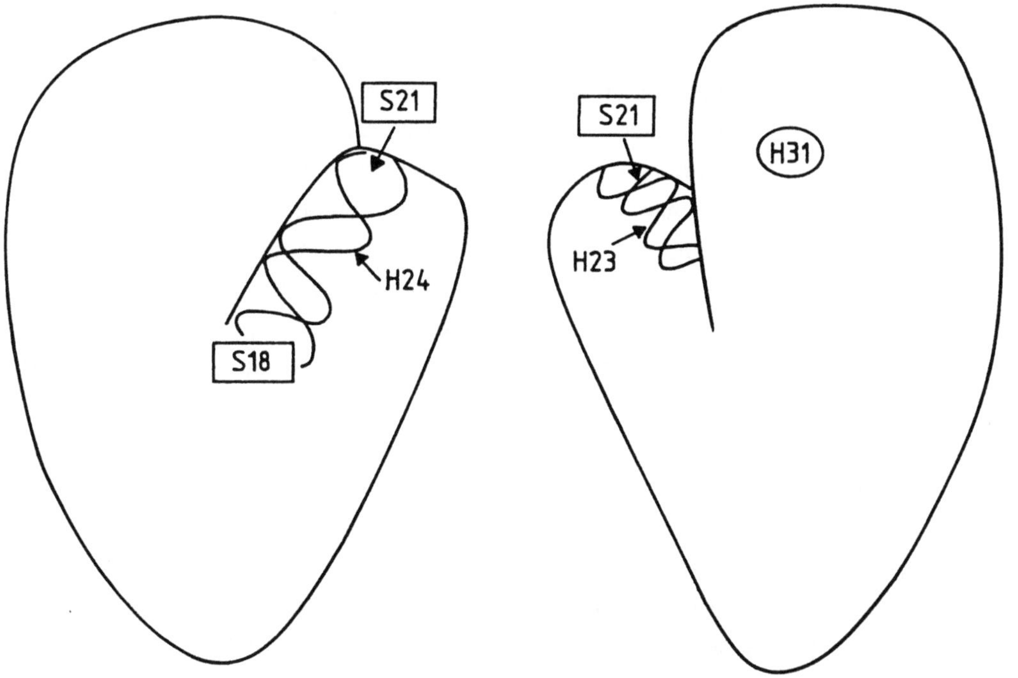

Figure 2.- Model of the 30S ribosomal subunit indicating the position of helices H23, H24 and H31 and of proteins S18 and S21 (Brimacombe et al, 1988).

On the other hand, pactamycin has also been shown to affect the reactivity of nucleotides G693 and C795 by chemical footprinting (J.Woodcock, Moazed and Noller; J.Egebjerg, referred in Cundliffe, 1990) . Interestingly, these nucleotides are also protected by edeine (Moazed and Noller, 1987a), another inhibitor of protein synthesis initiation , that, however, is reported to block a different step of the initiation process than pactamycin (Vazquez, 1979).

Taken together in the model of the 16S rRNA structure (Brimacombe et al., 1988) the different experimental data appear to be rather scattered at a first glance (fig 1). However, on closer analysis it is clear that two of the photolabeled proteins , S18 and S21 are found closely related to helices 23 and 24 and their corresponding loops, where nucleotides G693 and C795 are located.It seems therefore that the domain configurated by all these components is important for pactamycin binding. On the other hand, the methylated A964 is in the loop formed by helix 31, for which there are no data suggesting a close relation with the other pactamycin related elements. In fact, in the 30S subunit model constructed taking into consideration the rRNA secondary structure (Brimacombe et al. 1988) it has been recently confirmed by electron microscopy (Oakes and Lake, 1990), that helices 23 and 24 as well as proteins S21 and S18 form part of the lateral platform, a region (fig 2) which is well known to be involved in the initiation step of protein synthesis. According to this model ,however, helix 31 and the A964 loop are rather buried in the head of the subunit opposite the plataform. It is difficult to visualize, unless the models are basically wrong, how all these components can form part of the actual pactamycin binding site. It is possible, however, that the binding site is in the the plataform close to the subunit head, and methylation of A964 changes the conformation of that region, hindering the interaction of the drug.

The case of the pactamycin binding site is, as far as the agreement of the results of different techniques is concerned, in some ways an intermediate situation. Most of them can be shaped into a rather coherent picture while others, like the labelling of protein S4, which is located in the 30S subunit model realtively separated from the other pactamycin- related components, are difficult to fit in this scheme unless a rather elongated shape for the S4 is proposed.

One interesting facet of the proposed interpretation of the pactamycin data is that it assumes the existence of long distance effects of the drug interaction, accepting the fact that antibiotic resistant phenotypes can be due to cooperative effects of the ribosomal structure. Consequently, the ribosomal components, either nucleotides or proteins, affected in antibiotic resistant ribosomes, might not necessarely be part of the drug binding site. This important conclusion is sometimes disregarded when interpreting results from this type of experiments.

Macrolides

The macrolides form a relatively homogeneous group of compounds having a glycosylated lactone ring that inhibits protein synthesis by interacting with the ribosome. These antibiotics can be divided into two main groups on the basis of their biological activity (Vazquez, 1979). One group of macrolides is good inhibitor of peptide bond formation model reactions, while the other is inactive in these assays. The peptide bond inhibitors have a 16 atom lactone ring glycosylated by a disaccharide, while the non-inhibitors have a 14- atom lactone modified by monosaccharides. Spiramycin and erythromycin are the best -known members of each group.

In spite of their structural and functional differences, all macrolides have been shown to compete among themselves for binding to the ribosome, indicating the existance of closely related binding sites (Vazquez, 1979). The apparent contradiction between the functional and the binding results can be explained by the existence of different but overlapping binding sites for the two groups. Alternatively, both types of drugs could bind to the same site, perhaps through the common lactone ring, but the different structural differences in the drug molecules would result in different biological effects.

Chemical modification of the drug molecule was carried out in an attempt to understand the peculiar activity of macrolides. The results from these experiments suggest that, indeed, the size of the residues modifying the lactone ring affects the inhibition pattern of the drug (Arévalo et al. 1988, Ballesta and Lázaro,1990). Recently, we have been able to confirm these points using a number of derivatives of tylosin, a 16-atom lactone macrolide (Kirst et al. 1988), which have progressively shorter sugar chains glycosylating the lactone. As shown in figure 3 , as the sugar residue changes from a disaccharide to a monosaccharide, the pattern of inhibition changes as well , from the spiramycin (peptide bond inhibitor) to the erythromycin (peptide bond non-inhibitor) mode of action while the antibacterial activity remains unaffected. These results are compatible with, but do not prove the hypothesis of one binding site for all the macrolides, which , depending on the size of the lateral chain, will affect differently the ribosome activity. In fact ,it has been shown that erythromycin will inhibit the peptide bond reaction if the donor peptide is larger than five or six amino acids, supporting the idea that spiramyicn -type macrolides, carrying a disaccharaide, can reach further from the same binding site and therefore block the reaction with shorter donor peptides (Contreras and Vazquez, 1977).

EFFECT OF GLYCOSYLATING CHAIN ON MACROLIDES ACTIVITY

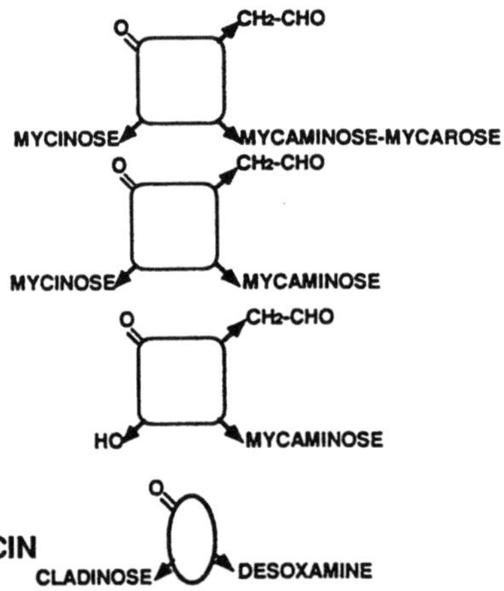

	ACTIVITY	
	Inhibition of Poly-Phe Synthesis at 1 μM %	Inhibition of Growth MIC μM
TYLOSIN	85.4	0.5
DESMYCOSIN	14.9	0.5
OMT	0	1.0
ERYTHROMYCIN	12.8	0.5

Figure 3.- Inhibitory activity of erythromycin, tylosin and some tylosin derivatives. 16 -atom and 14-atom lactone rings are represented respectively by squares and ovals in the structural schemes. *E. coli* cells and *E. coli* derived cell free systems were used for *in vivo* and *in vitro* tests

The identification of the structural components of the macrolide binding site will obviously help to decide which of the two binding models is correct. The data on the possible components of the macrolide binding site are relatively abundant.

Erythromycin and carbomycin (a 16 -atom ring macrolide) interaction has been studied by chemical footprinting by Moazed and Noller (1987b) who found that both drugs protect the nucleotides A_{2058}, A_{2059} and G_{2505}; in addition A_{2062} and A_{2451} are protected only by carbomycin, which is compatible with the idea that the 16 -atom ring macrolides have a larger binding site. These nucleotides are in the highly conserved region of domain V in the 23S rRNA that , as data from different sources have suggested , plays an important function in the peptidyl transferase center (Vester and Garrett, 1988, Steiner, 1988).

The analysis of the ribosomal components from macrolide producing organisms naturally resistant to the drugs, has confirmed the importance of the role domain V of the 23S rRNA plays in the interaction of macrolides (Cundliffe,1989). The only alteration found in the resistant ribosome was the methylation of A_{2058}. Mutations inducing resistance to erythromycin in mitochodria have been similarly mapped in nucleotides in positions equivalent to *E. coli* A_{2058} and C_{2611} (Sor and Fukuhara, 1982,1984). Moreover, mutations in the same positions of the *E.coli* 23S rRNA have been shown to induce resistance to erythromycin when introduced in multicopy plasmids (Sigmund et al. 1984,Ettayebi et al. 1985, Vester and Garret, 1987). Interestingly, all these mutations induce resistance not only to macrolides of both types but to the so called MLS antibiotics (macrolides, lincomycins, streptogramins). All these inhibitors act on the peptidyl transferase center and, therefore, these data strongly support the implication of domain V of the 23S rRNA in this ribosomal active center and favor the existence of a common mechanism for their inhibitory action . However, it must be indicated that these results, although compatible with a direct implication of these nucleotides on the binding site of all these antibiotics, do not prove that they form part of their actual binding site. The differences reported to exist in the mode of action of the various drugs (Vazquez, 1975) suggest that some specific components must exist in the interaction site of each drug. In fact, mutations inducing dissociated resistance to different drugs support the idea of the specificity and independence of the antibiotic binding sites. Thus, coming back to the macrolides, resistance to erythromycin has been reported to appear upon deletion of nucleotides 1219-1230 from domain II in the 23S rRNA (Douthwaite et al, 1985) and also when there are alterations of ribosomal proteins L4 and L22 (Wittman et al. 1973; Pardo and Rosset, 1977).

On the other hand, protein L16 was found to be fundamental in order to preserve the erythromycin binding capacity of protein -deficient ribosomal particles (Bernabeu et al, 1977). Protein L16 has been shown to induce an important conformational change in the ribosomal particle (Teraoka and Nierhaus, 1978) and ,therefore, it is possible that the effect on drug binding, as well as on other ribosomal activities (Bernabeu et al. 1977) is rather indirect.

Affinity labeling methods have also been utilized to study the macrolide binding site (Tejedor and Ballesta, 1985,1986; Arevalo et al. 1988). Protein L27 has been shown to be the main protein labeled by 16 -atom lactone macrolides, while protein L22 and to a lesser extent protein L15, are labeled by erythromycin. Although these data suggest the existence of different components in the two binding sites it has to be taken into account that the photoreactive groups are not the same, and they are not in the same relative position in the drug molecules in both cases.

Recently, photolabeling of the 23S rRNA using a derivative of erythromycin has been achieved in our laboratory. Although definitive location of the labeled nucleotide has not been carried out yet, the avialable results using RNAse H experiments clearly indicate that the label is in a fragment covering from nucleotide 2100 up to the 3'end, which is included in domain V but excludes A_{2058} (Lazaro and Ballesta, unpublished results).

When all the data on the macrolide binding site components are included in the 23S rRNA secondary structure model, it is clear that, in spite of the fact that information available on this rRNA is less complete than in the case of the 16s rRNA, the resulting picture (fig 4) is less coherent than in the case of pactamycin. The proteins that either by affinity labeling (L15, L22, L27) or by resistant mutant analysis (L22, L4) have been related to macrolide interaction have not been related in any way to domain V, where all the macrolide -binding -related nucleotides have mapped. Only protein L16 has been reported to interact with the 3' end fragment of the 23S rRNA (Zimmerman,1980) but , as we have indicated previously , the involvement of this protein in erythromycin binding has to be taken with some reservations. Protein L22, however, is reported to be somehow involved with domains I and II and precisely a deletion on the second one induces resistant to erythromycin (Douthwaite et al. 1985), suggesting the existence of at least a functional connection of this part of the ribosome structure with domain V.

A model for distribution of proteins in the 50S subunit has recently been published (Walleczek et al. 1989). According to the model, proteins L27, L22 are located not far from each other in a central part of the structure close to but not right in the peptidyl transferase center . This location is expected if a unique binding site for all the macrolides is present in the ribosome. In addition, this site has to be displaced from the peptide bond formation center, since, otherwise, the activity should directly be inhibited by all the macrolides, including erythromycin. From a displaced interaction site, the 16 -atom ring macrolides could directly reach the active center due to their longer disaccharide chain, while erythromycin, having a monosaccharide requires that the growing peptide should reach a certain length to be able to block the process. Based on these considerations, protein L22 was located (Arevalo et al. 1988) precisely in the same position where it has been placed in a recent 50S subunit model taking in account other experimental data (Wallaczek, et al. 1989).

It seems clear that taking all the experimental data together, the situation is rather confusing. At the present moment it is dificult to reconcile the protein data, either from affinity labeling or from resistant mutant analysis, with most of the rRNA data. Nor have the structural data helped to resolve the dilemma of the presence of one or more independent binding sites for the two macrolide groups, although functional results seem to suggest the first alternative. It is obvious then, that additional information is required before a more clear -cut picture of the macrolide interaction site is obtained. This may eventually come from the development of new affinity labeling derivatives.

FINAL COMMENTS

It is obvious, considering the cases presented here, that total agreement of the results obtained from different methodological approaches used to study the structure of the binding site is frequently not to be found. Nor is it always possible to fit all the available data into (existing) ribosome models to obtain a coherent proposal for the binding site structure . This lack of accord can be partially due to the still fragmentary knowledge of the ribosomal structure; however,it is also probable that in some cases, indirect effects of the drug interaction rather than the actual components of the binding site are recorded. If this possibility is not accepted, some data that correctly interpreted could provide important information on the antibiotic mode of action, will be dismissed as "artefacts", most of the time based on purely subjective interpretations. In this context, it is perhaps important to note now, that RNA is assuming a principal role in the ribosome function and more precisely in the drug binding interaction (Cundliffe, 1989). The large rRNA molecules with complex secondary and tertiary structures are probably more susceptible to allosteric effects than the rather small ribosomal proteins, and special precaution should be exercised when interpreting results that implicate rRNA modifications.

The three cases commented here in detail illustrate how much potential information a detailed knowledge of the antibiotic binding sites would provide in order to define more precisely a ribosomal active center. Examples of this would be the elongation factors interaction center in the case of thiostrepton, or establishing the existence of allosteric interactions between different parts of the ribosome particle as was discussed for pactamycin. Also, the data on the macrolides,although the actual structure of their binding site is still far from clear, have been very useful in determining ribosome components implicated in the peptidyl transferase center.

Future developments in this field will explore in more detail the possibilities of the different technical approaches . In the case of affinity labeling, characterization of the labeled components will probably be required and, in the case of proteins, trying to identify the amino acids implicated in the interaction will be essential.

ACKNOWLEDGEMENTS

I thank Dr. H.A. Kirst from Lilly Research Laboratories, Indianapolis, USA, for the gift of tylosine derivatives.

Part of the work described has been supported by institutional grant to Centro de Biologia Molecular from Fundacion Ramon Areces and by grant PB0450 from Direccion General de Investigacion Cientifica y Tecnica (Spain).

REFERENCES

Arevalo, M.A., Tejedor, F., Polo, F. and Ballesta, J.P.G. (1988) Protein components of the erythromycin binding site in bacterial ribosomes J. Biol. Chem. 263: 58-63.
Ballesta, J.P.G. and Lazaro, E. (1990) Peptidyl transferase inhibitors: Structure - activity relationship analysis by chemical modification p502-510 In The Ribosome. Structure, Function and Evolution (Hill, W.E., Dahlberg, A., Garret, R.A., Moore, P.B., Schelissinger, D. and Warner, J.R. eds) American Society for Microbiology, Washington, D.C.
Bernabeu, C.,Vazquez, D, and Ballesta, J.P.G. (1977) The involvement of protein L16 on ribosomal peptidyl transferase activity. Eur. J. Biochem. 79: 469-472.
Brimacombe, R., Atmadja, J.,Stiege W. and Schüler, D. (1988) A detailed model of the three-dimensional structure of *E. coli* 16S ribosomal RNA *in situ* in the 30S subunit. J. Mol. Biol. 199: 115-136.
Calcutt, M.J. and Cundliffe, E. (1990) Resistance to pactamycin in clones of *Streptomyces liovidans* containing DNA from pactamycin-producing *Streptomyces pactum* . Gene, 93, 85-89.
Contreras, A. and Vazquez, D. (1977) Cooperative and antagonistic interactions of peptidyl-tRNA and antibiotics. Eur. J. Biochem. 74, 538-547.
Cooperman, B.S. (1980) Functional sites on E. coli ribosomes as defined by affinity labeling p 531-554 In Ribosomes. Structure, Function and Genetics (Chamblis, G., Craven, G.R., Davies, J., Davis K., Kahan, L. and Nomura, M. eds.) pp531-554. University Park Press, Baltimore, USA
Cundliffe. E. (1990) Recognition sites for antibiotics within rRNA p 479-490 In The Ribosome. Structure, Function and Evolution (Hill, W.E., Dahlberg, A., Garret, R.A., Moore, P.B., Schelissinger, D. and Warner, J.R. eds) American Society for Microbiology, Washington, D.C.
Cundliffe, E. (1989) How antibiotic-producing organisnms avoid suicide Ann. Rev. Microbiol. 43: 207-233.
Cundliffe, E. and Thompson, J. (1979) Ribosome methylation and resistance to thiostrepton. Nature 278, 859-861.
Douthwaite, S., Prince,J.B. and Noller, H.F. (1985) Evidence for functional interaction between domains II and V of 23S ribosomal RNA from an erythromycin-resistant mutant Proc. Nat. Acad. Sci. USA 82: 8330-8334.
Egebjerg, J., Douthwaite, S. and Garret, R.A. (1989) Antibiotic interactions at the GTPase-associated centre within *Escherichia coli* 23S rRNA EMBO J. 8: 607-611.
Ettayebi, M., Prasad, S.M. and Morgan, E.A. (1985) Chloramphenicol-erythromycin resistance mutations in a 23S rRNA gene of *Escherichia coli* J. Bacteriol. 162: 551-557.
Gale , E.F., Cundliffe, E., Reynolds, P.E., Richmond, M.H. and Waring, M.J. (1981) The molecular basis of antibiotic action, John Wiley and Sons, London and New York.
Girshovich, A.S., Bochkareva, E.S. and Vasiliev V.D.(1986) Localization of elongation factor Tu on the ribosome. FEBS Lett. 197: 192-198.

Herr, W., Chapman, N.M. and Noller, H.F. (1979) Mechanism of ribosomal subunit association: Discrimination of specific sites in 16S RNA essential for association activity. J. Mol. Biol. 130: 433-449.
Hummel, H. and Böck, A. (1987) Thiostrepton resistance mutations in the gene for 232S ribosomal RNA on halobacteria Biochimie 69: 857-861.
Jimenez, A. and Vazquez, D. (1975) Quantitative binding of antibiotics to ribosomes from a yeast mutant altered on the Peptidyl Transferase center Eur. J. Biochem. 54, 483-492.
Kirst, H.A., toth, J.E., Debono, M., Willard, K.E., Truedell, B.A., Ott, J.L., Counter, F.T., Felty-Duckworth, A.M. and Pekarek, R.S. (1988) Synthesis and evaluation of tylosin-related macrolides modified at the aldehyde function: a new series of orally effective antibiotics J. Med. Chem. 31, 1631-1641.
Moazed, D. and Noller, H.F. (1987a) Interaction of antibiotics with functional sites in 16S ribosomal RNA Nature 327: 389-394.
Moazed D. and Noller, H.F. (1987b) Chloramphenicol, erythromycin, carbomycin, and vernamycin B protect overlaping sites in the peptidyl transferase region of 23S ribosomal RNA Biochimie 69: 879-884.
Oakes , M.I. and Lake J.A. (1990) DNA-hybridization electron microscopy: localization of five regions of 16S rRNA on the surface of the 30S ribosomal subunit. J. Mol. Biol. 211: 897-906.
Oakes,M.I.,Kahan,L. and Lake, J.A. (1990) DNA-hybridization electron microscopy: tertiary structure of 16S rRNA. J. Mol. Biol. 211: 907-918.
Pardo, D. and Rosset, R. (1977) Genetic dtudies of erythromycin resistant mutants of *Escherichia coli* Mol. Gen. Genet. 156,267-271.
Raue, H.A., Klootwijk, J. and Musters W. (1988) Evolutionary conservation of structure anf function of high molecular weight ribosomal RNA Prog. Biophys. Molec. Biol. 51: 77-129.
Sigmund,C.D.,Ettayebi, M. and Morgan, E.A. (1984) Antibiotic resistance mutations in 16S and 23S ribosomal RNA genes of *Escherichia coli* Nucleic Acids Res. 12, 4653-4663.
Sor, F.,and Fukuhara, H. (1982) Identification of two erythromycin resistance mutations in the mitochondrial gene coding for the large ribosomal RNA in yeast Nucleic Acids Res. 10: 6571-6577.
Sor, F. and Fukuhara, H. (1984) Erythromycin and spiramycin resistance mutations of yeast mitochondria: nature of the rib2 locus in the large ribosomal RNA Nucleic Acids Res 12: 8313-8318.
Schüler, D. and Brimacombe R. (1988) The E. coli 30S ribosomal subunit; an optimized three-dimensional fit between the ribosomal proteins and the 16S RNA. EMBO J. 7: 1509-1513
Steiner, G., Kuechler,E. and Barta, A. (1988) Photoaffinity labeling at the peptidyl transferase center reveals two different positions for the A- and P- sites in domain V of 23S rRNA EMBO J. 7, 3949-3955.
Synetos, D., Amils, R. and Ballesta, J.P.G. (1986) Photolabeling of protein components in the pactamycin binding site of rat liver ribosomes Biochim. Biophys. Acta 868, 249-253.
Tapprich, W.E., Goss, D.J. and Dahlberg, A.E. (1989) Mutation at position 791 in *Escherichia coli* 16S ribosomal RNA affects processes involved in the initiation of protein syntyhesis Proc. Natl. Acad. Sci. USA 86: 4927-4931.

Tejedor, F. and Ballesta, J.P.G. (1985) Ribosome structure: binding site of macrolides studied by photoaffinity labeling. Biochemsitry 24: 467-472.
Tejedor, F., Amils, R. and Ballesta, J.P.G. (1985) Photoaffinity labeling of pactamycin binding site on eubacterial ribosomes Biochemistry, 24: 3667-3672.
Tejedor, F. and Ballesta, J.P.G. (1986) Reaction of some macrolide antibiotics with the ribosome. Labeling of the binding site components Biochemistry 25,7725-7731.
Tejedor, F., Amils, R. and Ballesta. J.P.G. (1987) Pactamycin binding site on archaebacterial and eukaryotic ribosomes Biochemistry 26:652-656.
Teraoka, H. and Nierhaus, K. (1978) Protein L16 induces a conformational change when incorporated into an L16-deficient core derived from *Escherichia coli* ribosomes. FEBS Lett. 88: 223-227.
Thompson, J., Cundliffe, E. and Stark, M. (1979) Binding of thiostrepton to a complex of 23S rRNA with ribosomal protein L11. Eur. J. Biochem. 98: 261-265.
Thompson, J., Schmidt, F. and Cundliffe, E. (1982) Site of action of a ribosomal RNA methylase conferring resistance to thiostrepton. J. Biol. Chem. 257: 7915-7917.
Thompson, J., Cundliffe, E. and Dahlberg, A.E. (1988) Site-directed mutagenesis of *Escherichia coli* 23S ribosomal RNA at position 1067 within the GTPase hydrolysis center. J. Mol. Biol. 203, 457-465.
Vester, B. and Garrett, R.A. (1987) A plasmid -coded and site-directed mutation in *Escherichia coli* 23S RNA that confers resistance to erythromycin: implications for the mechanism of action of erythromycin. Biochimie 69, 891-900.
Vester, B. and Garrett, R.A. (1988) The importance of highly conserved nucleotides in the binding region of chloramphenicol at the peptidyl transferase centre of *Escherichia coli* 23S ribosomal RNA. EMBO J. 7, 3577-3587.
Vazquez, D. (1979) Inhibitors of protein synthesis. Mol. Biol. Biochem.Biophys. 30,1-312.
Walleczek, J., Schüler,D., Stöffler-Meilicke, M., Brimacombe, R. and Stöffler, G. (1988) A model for the spatial arrangement of the proteins in the large subunit of the *Escherichia coli* ribosome. EMBO J. 7, 3571-3576.
Wittmann, H.G., Stöffler, G., Apirion, D., Rosen,L., Tanaka, K., Tamaki, M.Tanaka,R.,Dekio,S., Otaka, E. and Osawa, S. (1973) Biochemical amd genetic studies on two different types of erythromycin resistant mutants of *Escherichia coli* with altered ribosomal proteins. Mol.Gen Genet. 127, 175-189.
Zimmermann, R. (1980) In Ribosomes, Structure, Function and Genetics (Chamblis, G., Craven G.R., Davies, J, Davis, K. Kahan, L. and Nomura, M., eds.) pp 135-169. University Park Press, Baltimore, USA

RECENT PROGRESS ON UNDERSTANDING THE INITIATION OF TRANSLATION IN THE CHLOROPLASTS OF *EUGLENA GRACILIS*

W. B. Roney, L. Ma, C.-C. Wang and L. L. Spremulli
Department of Chemistry
University of North Carolina
Chapel Hill, NC 27599-3290, USA

The mechanism of protein biosynthesis in chloroplasts is currently not well understood. However, in general, the process of translation in this organelle is thought to resemble that in prokaryotes more closely than that in the eukaryotic cell cytoplasm. For example, chloroplast ribosomes are 70 S in size and initiation occurs with formylmethionine (Kozak, 1983). In bacteria, the initiation of translation requires three auxiliary factors. IF-2 promotes the binding of fMet-tRNA to 30 S ribosomal subunits. IF-3 facilitates ribosome dissociation and is thought to proofread the initiation complex. IF-1 facilitates the action of both IF-2 and IF-3 (Gualerzi and Pon, 1990). In contrast, the eukaryotic cytoplasmic system requires about 10 factors for the formation of the completed initiation complex (Pain, 1986). Several of these are involved in the binding of Met-tRNA to the ribosome while others appear to be required for the interaction of the 40 S subunit with the 5' cap structure on the message.

Little information is available on the nucleotide sequences that specify the start site for protein synthesis on chloroplast mRNAs. In prokaryotes, the selection of the start site for translation is defined in part by the presence of a sequence (the Shine/Dalgarno sequence) that is complementary to a stretch of nucleotides near the 3' end of the small subunit rRNA. In contrast, cytoplasmic 40 S subunits interact with the cap structure on the mRNA and, in the presence of appropriate initiation factors, scan the mRNA for the first AUG codon where initiation generally occurs (Kozak, 1983). An examination of putative start sites on chloroplast mRNAs indicates that some of them contain potential Shine/Dalgarno sequences while others do not have a corresponding sequence (Ruf and Kössel, 1988; Bonham-Smith and Bourque, 1989). These observations suggest that the recognition of the start site on certain chloroplast mRNAs may resemble the bacterial mechanism while initiation on other chloroplast mRNAs will use another, as yet poorly understood, mechanism.

In the present paper, we report the identification, purification and partial characterization of two chloroplast translational initiation factors from *Euglena gracilis*. One of these factors ($IF\text{-}2_{chl}$) is responsible for binding fMet-tRNA to chloroplast 30 S subunits while the other (IF-3_{chl}) promotes initiation complex formation in the presence of IF-2_{chl}. In addition, we describe the formation of initiation complexes on chloroplast 30 S subunits with natural mRNAs and our initial analysis of mRNA sequences important for recognition of the start sites on *E. gracilis* chloroplast mRNAs.

Materials and Methods

Purification of Chloroplast IF-2 and IF-3: *E. gracilis* was grown and a postribosomal supernatant was prepared as described (Ma and Spremulli, 1990). The postribosomal supernatant was used for the preparation of both IF-2_{chl} and IF-3_{chl}. Multiple forms of IF-2_{chl} were prepared as described by Ma and Spremulli (1990). IF-3_{chl} was partially purified as described previously (Kraus and Spremulli, 1986) and further purified by chromatography on hydroxyapatite and by high pressure liquid chromatography on sulfopropyl-, DEAE- and heparin-resins (C.-C. Wang, manuscript in preparation). Chloroplast 30 S ribosomal subunits were prepared as described (Graves and Spremulli, 1983).

Initiation Complex Assays: The activities of IF-2_{chl} and IF-3_{chl} were monitored as described previously (Kraus and Spremulli, 1986; Ma and Spremulli, 1990). The formation of initiation complexes with natural mRNAs were carried out by the procedure of Wang *et al.* (1989).

Plasmids and mRNAs: A plasmid (designated pRbcN) carrying the 5' untranslated leader region of the *rbcL* gene fused in frame to an internal coding region of the neomycin phosphotransferase gene was prepared as described (Wang *et al.*, 1989). This construct contains a T7 polymerase promoter and allows the synthesis of mRNA *in vitro* that carries the complete 5' untranslated leader of the *rbcL* gene. Unique XbaI sites were introduced into the extreme 5' and 3' ends of the 5' untranslated leader region of the *rbcL* gene using oligonucleotide-directed mutagenesis generating a plasmid designated the Xba mutant. A deletion of all but 10 nucleotides of the 5' untranslated leader region was prepared by cutting this construct with XbaI, gel purification of the large fragment followed by religation (Maniatis *et al.*, 1982). *In vitro* transcription reactions and mRNA purification were carried out as described (Wang *et al.*, 1989).

Results

Purification and Characterization of IF-2$_{chl}$: IF-2$_{chl}$ promotes the binding of fMet-tRNA to chloroplast 30 S ribosomal subunits. Early work (Gold and Spremulli, 1985) indicated that it is present in the postribosomal supernatant of whole cell extracts of *E. gracilis*. This factor may be significantly enriched by chromatography on phosphocellulose resins. When partially purified preparations of IF-2$_{chl}$ are subjected to chromatography on Sephacryl S-300 resins, this activity can be resolved into multiple high molecular mass forms ranging from 200 kDa to over 700 kDa (Ma and Spremulli, 1990). This observation is somewhat surprising since the corresponding bacterial and cytoplasmic factors are much smaller and more uniform in size. The complex forms of IF-2$_{chl}$ have been further purified by a combination of conventional and high pressure liquid chromatography as indicated in Fig. 1.

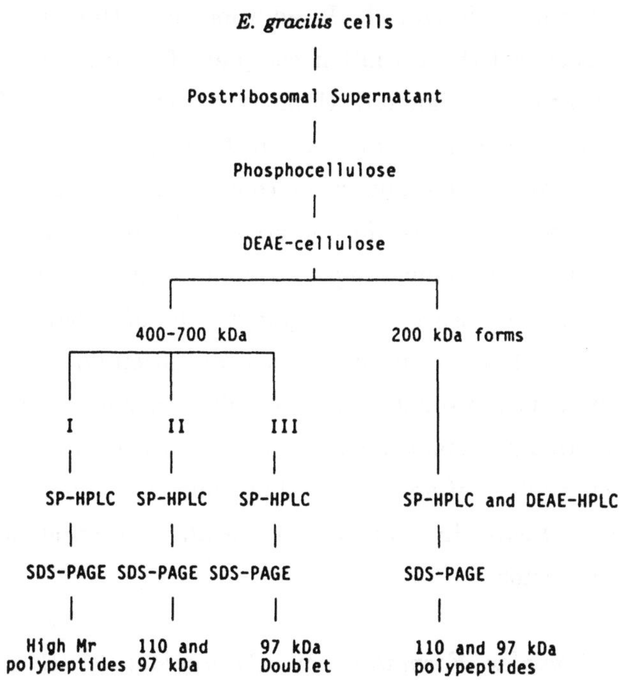

Fig. 1: Flow diagram for the isolation of the complex forms of IF-2$_{chl}$.

Analysis of the purified forms of IF-2$_{chl}$ on SDS-polyacrylamide gels indicates that there are several polypeptide components present. One form of

IF-2_{chl} appears to be composed of polypeptides of about 200 kDa in size which may be present in dimer or tetramer aggregates. Other forms appear to consist of complexes of polypeptides of about 110 and 97 kDa which may be present in dimeric and tetrameric aggregates under native conditions.

A number of the properties of IF-2_{chl} have been investigated. This factor promotes fMet-tRNA binding to 30 S subunits in a message-dependent reaction. Initiation complex formation by IF-2_{chl} requires the presence of GTP. Similar levels of binding are obtained when GTP is replaced by a non-hydrolyzable analog suggesting that IF-2_{chl} is acting stoichiometrically rather than catalytically under the conditions used. GTP and fMet-tRNA appear to be present in a 1:1 stoichiometry in the 30 S initiation complexes formed. None of the forms of IF-2_{chl} detected is active on *E. coli* ribosomes.

Purification and Characterization of IF-3_{chl}: In *E. coli* IF-3 promotes initiation complex formation by binding to the 30 S subunit and preventing its reassociation with the 50 S subunit. In addition, it is thought to stabilize IF-2 binding and to proofread the initiation complex (Gualerzi and Pon, 1990). We have detected a factor designated IF-3_{chl} that will replace *E. coli* IF-3 in promoting initiation complex formation on *E. coli* ribosomes. A purification scheme has been developed for this factor that results in the resolution of three forms of IF-3_{chl}. Two of these forms appear to be composed of a single polypeptide having a molecular weight of about 40,000 while the third form appears to be a monomer with a molecular weight of about 35,000.

IF-3_{chl} will stimulate initiation complex formation on chloroplast or *E. coli* ribosomes and will also promote ribosome dissociation. In these respects it appears to be functionally analogous to its prokaryotic counterpart. However, *E. coli* IF-3 is a basic protein of about 22,000 in molecular weight (Hershey, 1987). In contrast, the *E. gracilis* IF-3_{chl} is a rather acidic protein about twice the size of the prokaryotic factor.

Initiation Complex Formation on Chloroplast 30 S Subunits in the Presence of Natural mRNA: At the present time there is little information on the nucleotide sequences which specify the start signal for protein synthesis on chloroplast mRNAs. However, an examination of the putative translation start sites on *E. gracilis* chloroplast mRNAs indicates that many of them do not contain the Shine/Dalgarno sequences which are so prevalent in the translation start sites in prokaryotic mRNAs. In an effort to examine start signals on chloroplast mRNAs, we have developed a system that allows the assembly of an initiation complex on chloroplast 30 S subunits (Wang *et al.*, 1989). For these experiments, the 5' untranslated leader region and exon 1 of the *E. gracilis rbcL* gene were cloned in frame to a longer open reading frame behind the T7 transcriptional promoter (this construct is designated *pRbcN*). Messenger RNA prepared from this plasmid will actively promote fMet-tRNA binding to chloroplast 30 S subunits. As indicated in Fig. 2, both IF-2_{chl} and IF-3_{chl} are important for the formation of the initiation complex with the *rbcN* mRNA. Removal of IF-2_{chl} from reaction mixtures results in the loss of essentially all of the activity in initiation complex formation while omission of IF-3_{chl} results in a 3- to 4-fold reduction in the amount of fMet-tRNA bound to chloroplast 30 S subunits.

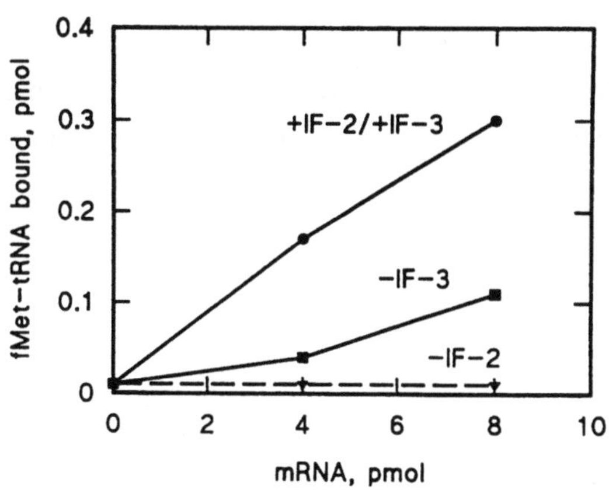

Fig. 2. Importance of IF-2_{chl} and IF-3_{chl} for the binding of fMet-tRNA to chloroplast 30 S subunits in the presence of the *rbcN* mRNA. Initiation complex formation in the presence of both IF-2_{chl} and IF-3_{chl} (-●-); Complex formation in the absence of IF-3_{chl} (-■-); fMet-tRNA binding in the absence of IF-2_{chl} (-▾-). All reaction mixtures were prepared as described by Wang *et al.* (1989) and contained *E. coli* IF-1.

We have begun to examine the role of the 5' untranslated leader sequences in the *rbcL* gene (present in the *rbcN* construct) in specifying the initiation of translation in *E. gracilis* chloroplasts. For this work, we wished to delete the majority of the 5' leader sequence in order to ascertain whether the information required to designate a specific AUG as the initiation codon was indeed located in the 5' untranslated region of the mRNA. The 5' leader of the *rbcL* gene is extremely A/T rich and does not contain any convenient restriction enzyme sites (Fig. 3). To facilitate the construction of appropriate deletions, we introduced two XbaI sites (Fig. 3) between the transcription start site and the AUG codon using site-directed mutagenesis (designated the Xba mutant).

GTATAAATAACTGTAAAGTGATTTTTCGAATATAATTTTTATTTTATTATTAAAATATG ...	*RbcN*
GTCTAGATAACTGTAAAGTGATTTTTCGAATATAATTTTTATTTTATTATCTAGAAATATG...	*Xba mutant*
GTCTAGAAATATG...	*5' deletion*

Fig. 3: The 5' untranslated leader regions of the wild type *rbcL* gene, the mutant containing 2 unique XbaI sites (underlined) and the mutant which has only 10 nucleotides 5' to the start AUG codon (underlined).

The preparation of the XbaI sites in the untranslated leader region required the introduction of a G and a C residue near the start codon resulting in a lower % A/U content near the start AUG. However, as indicated in Fig. 4, this small change in the mRNA sequence near the start codon had essentially no effect on the ability of the resulting mRNA to function in initiation complex formation.

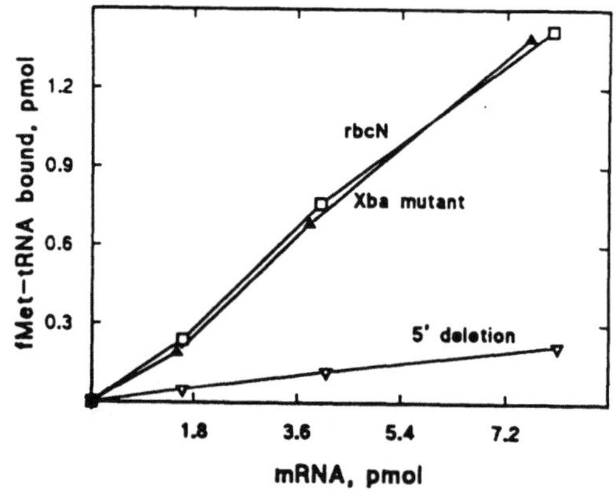

Fig. 4: Initiation complex formation on mRNAs with various 5' untranslated leader regions. Initiation complexes were formed (Wang *et al.*, 1989) using *E. coli* initiation factors and the mRNA carrying the wild type leader for the *rbcL* gene (-□-); the 5' leader from the XbaI mutant (-▲-) or the deletion mutant that contains only 10 nucleotides 5' to the translation start site (-▽-).

Deletion of the XbaI fragment in the 5' leader of the *rbcL* gene results in the synthesis of a message that contains only 10 nucleotides prior to the AUG start codon (Fig. 3). The resulting mRNA was essentially inactive in promoting fMet-tRNA binding to chloroplast 30 S subunits (Fig. 4). This observation suggests that information essential for specifying the start site on the *rbcL* mRNA is located in the 5' untranslated region and lays the foundation for a more detailed analysis of the nucleotide signals involved in initiation complex formation in *E. gracilis* chloroplasts.

Discussion

Two chloroplast protein synthesis initiation factors have been identified and purified from *E. gracilis*. Both of these factors are physically quite different from the corresponding bacterial factors. In *E. coli*, IF-2 is present in two forms having molecular weights of 97 kDa and 80 kDa (Hershey, 1987). In the eukaryotic cell cytoplasm, eIF-2 is a trimeric protein with a native molecular weight of about 140 kDa (Pain, 1986). In our work, we have observed that $IF\text{-}2_{chl}$ from *E. gracilis* occurs in high molecular weight complexes ranging from 200 kDa to over 700 kDa. The polypeptide components present appear to be primarily around 110 and 97 kDa. Thus, the subunits of $IF\text{-}2_{chl}$ are similar in size to the bacterial IF-2 but the factor appears to form complex aggregates and the physiologically active form of the remains to be determined. $IF\text{-}3_{chl}$ is also physically unusual compared to the bacterial factor. It is over twice the size of *E. coli* IF-3 and is an acidic rather than a basic protein. There is no factor directly equivalent to IF-3 in the eukaryotic cytoplasmic system.

Regardless of their unusual physical properties, the chloroplast initiation factors appear to carry out basically the same functions as do the bacterial factors. $IF\text{-}2_{chl}$ promotes the binding of fMet-tRNA to chloroplast small subunits while $IF\text{-}3_{chl}$ promotes initiation complex formation in the presence of $IF\text{-}2_{chl}$. However, $IF\text{-}2_{chl}$ is not active on bacterial ribosomes and it remains to be seen whether it functions mechanistically in an analogous manner. In addition, the mechanism of action of $IF\text{-}3_{chl}$ remains to be examined in detail. Although this factor will replace *E. coli* IF-3 in initiation complex formation, it is unclear whether this effect is manifested in a manner equivalent to that observed with the bacterial factor.

The interaction between the ribosome and the start signal on the mRNA during the initiation of translation is an intriguing process. In prokaryotes most of the mRNAs contain Shine/Dalgarno sequences which facilitate the selection of the correct AUG codon by base pairing with a polypyrimidine sequence near the 3' end of the 16 S rRNA. The 3' end of the small subunit rRNA from *E. gracilis* chloroplasts is rather unusual (Steege *et al.*, 1982). It contains a polypyrimidine sequence CUCCC which differs from the CCUCC sequence found most commonly in prokaryotes and in the 16 S rRNAs from other chloroplasts. The polypyrimidine tract in the *E. gracilis* rRNA actually terminates the chain rather than being located several nucleotides from the end. We have examined known *E. gracilis* chloroplast translation start signals for the presence of potential Shine/Dalgarno interactions within 20 nucleotides upstream of the AUG start codon (Wang *et al.*, 1989) and have observed that many of these mRNAs do not contain conventional Shine/Dalgarno sequences. For example, note the 5' untranslated leader region of the *rbcL* gene (Fig. 3). We have proposed that the translational machinery in *E. gracilis* chloroplasts has two potential mechanisms available for the recognition of translational start signals. One class of start sites has a potential Shine/Dalgarno sequence. In these mRNAs, the A/T content in the region up to -20 from the translation start site is often less than 90 % and the interaction between the 16 S rRNA and the mRNA may be important to promote initiation complex formation. In the second class of initiation sites, no Shine/Dalgarno sequence is present. In these mRNAs, the A/T content of the ribosome bound region 5' to the start AUG is very high (greater than 90 %) and this portion of the mRNA would be quite unstructured. These start sites may, therefore, be readily accessible to the 30 S subunit allowing initiation in the absence of a Shine/Dalgarno interaction. It should be noted that *E. coli* 30 S subunits are thought to interact preferentially with single-stranded stretches of mRNAs during the search for the translational start signal on bacterial mRNAs (de Smit and van Duin, 1990).

The observation that a message containing only 10 residues 5' to the AUG does not function effectively in initiation indicates that nucleotides 5' to the start codon play an important role in specifying the translation start site for the *rbcL* gene. There are two interpretations for this observation. First, it is reasonable to suppose that there is important sequence or structural information present in this region of the mRNA that is recognized by the initiating 30 S ribosomal subunit. The deletion could, therefore, have removed essential information from the mRNA. Alternatively, one might imagine that the chloroplast 30 S subunit may require a sequence of greater than 10 nucleotides 5' to the start AUG in order to bind tightly to the mRNA. In prokaryotes (De Smit and van Duin, 1990), it has been shown that the small subunit generally covers about 20 nucleotides 5' to the AUG codon in the initiation complex. It is, therefore, possible that the presence of only 10 nucleotides 5' to the AUG may prevent stable ribosome binding to the mRNA. Further detailed analysis of the translational start sites on *E. gracilis* chloroplast mRNAs is currently underway.

References

Bonham-Smith, PC and Bourque DP (1989) Translation of chloroplast-encoded mRNA: potential initiation and termination signals. *Nuc. Acids Res.*, **17**, 2057-2080

De Smit MH and van Duin J (1990) Control of prokaryotic translational initiation by mRNA secondary structure. *Prog. in Nucleic Acid Res. and Mol. Biol.*, **38**, 1-35

Gold JC and Spremulli LL (1985) *Euglena gracilis* chloroplast initiation factor 2 (IF-2_{chl}): Identification and initial characterization. *J. Biol. Chem.*, **260**, 14897-14900

Graves MC and Spremulli LL (1983) Activity of *Euglena gracilis* chloroplast ribosomes with prokaryotic and eukaryotic initiation factors. *Arch. Biochem. Biophys.*, **222**, 192-199

Gualerzi CO and Pon CL (1990) Initiation of mRNA translation in prokaryotes. *Biochemistry*, **29**, 5881-5889

Hershey, JWB (1987) Protein Synthesis in *Escherichia coli and Salmonella typhimurium* (Neidhardt F, ed) ASM Press

Kozak M (1983) Comparison of initiation of protein synthesis in prokaryotes, eukaryotes and organelles. *Micro. Rev.*, **47**, 1-45

Kraus, BK and Spremulli LL (1986) Chloroplast initiation factor 3 from *Euglena gracilis*: Identification and initial characterization, *J. Biol. Chem.*, **261**, 4781-4784

Ma L and Spremulli L (1990) Identification and Characterization of large, complex forms of chloroplast translational initiation factor 2 from *Euglena gracilis*. *J. Biol. Chem.*, **265**, 13560-13565

Maniatis T, Fritsch EF and Sambrook J (1982) *Molecular Cloning: A Laboratory Manual*, Cold Spring Harbor Laboratory

Pain V (1986) Initiation of protein synthesis in mammalian cells. *Biochem. J.*, **235**, 625-637

Ruf M and Kössel H (1988) Occurrence and spacing of ribosome recognition sites in mRNAs of chloroplasts from higher plants. *FEBS Lett.*, **240**, 41-44

Steege DA, Graves MC and Spremulli LL (1982) Euglena gracilis chloroplast small subunit rRNA: Sequence and base pairing potential of the 3' terminus, cleavage by colicin E3. *J. Biol. Chem.*, **257**, 10430-10439

Wang CC, Roney WB, Alston RL and Spremulli LL (1989) Initiation complex formation on *Euglena* chloroplast 30 S subunits in the presence of natural mRNAs. *Nuc. Acids Res.*, **17**, 9735-9747

THE EFFECTS OF CHLOROPHYLL PHOTOOXIDATION ON NUCLEAR-ENCODED PLASTID RIBOSOMAL PROTEIN mRNAs IN NORFLURAZON-TREATED PEA SEEDLINGS

J. Stephen Gantt, Anupma Gupta, Michael D. Thompson
Department of Plant Biology
University of Minnesota
1445 Gortner Ave.
St. Paul, Minnesota 55108
USA

INTRODUCTION

The development of functional chloroplasts from etioplasts or proplastids requires the expression of both nuclear and plastid genomes. Transcripts from genes located in the nucleus that encode proteins destined for the plastid are translated on cytosolic 80S ribosomes. To localize these proteins to the plastid these polypeptides contain an amino terminal transit sequence. The proteins are transported into the plastid where they are processed to their mature size through loss of their transit sequences. Messenger RNAs derived from plastid DNA are translated in the plastid compartment on 70S ribosomes. To form functional complexes, many of the plastid-encoded proteins associate with nuclear-encoded proteins and often all of these proteins accumulate in a coordinate manner. For example, the plastid ribosome is composed of equal numbers of approximately 55 different proteins, one-third of which are encoded in the plastid itself and the remaining two-thirds are encoded in the nucleus. The mechanisms responsible for the coordinate expression of genes located in different compartments are unknown.

In an effort to address the question of how genes located in different compartments are coordinately expressed, we have used a herbicide, norflurazon, that alters chloroplast development and have tested the effects that this herbicide has on the expression of the nuclear-encoded plastid ribosomal protein genes. Plastid derived effects on expression of nuclear-

NATO ASI Series, Vol. H 55
The Translational Apparatus of Photosynthetic Organelles
Edited by R. Mache et al.

encoded plastid genes have previously been demonstrated through the use of carotenoid-deficient mutants and/or norflurazon, which blocks carotenoid biosynthesis (Sagar et al. 1988; Batschauer et al. 1986; Mayfield and Taylor 1984). When mutant or herbicide-treated plants are exposed to bright light, chlorophyll is photooxidized due to the absence of carotenoids and little internal chloroplast membrane structure develops (Anderson and Robertson 1960). Additionally, plastid ribosomes and many other plastid proteins fail to accumulate (Reiss et al. 1983). In the carotenoid-deficient mutants and in leafs treated with norflurazon, many cytosolic mRNAs encoding plastid proteins do not increase to their normal light-induced levels or are not maintained at their normally high amounts (Burgess and Taylor, 1987). For example, the abundance of light-harvesting chlorophyll a/b protein (LHCP) mRNA is dramatically increased when normal barley seedlings are exposed to light, but treatment of the seedlings with norflurazon prior to their exposure to light blocks this light-induced increase in LHCP mRNA (Batschauer et al. 1986). The levels of mRNAs encoding cytosolic proteins are typically not affected in the mutants or in the plants treated with norflurazon. The loss of chlorophyll does not seem to be responsible for the altered gene expression since mutants that lack chlorophyll accumulate normal levels of LHCP mRNA (Harpster et al. 1984). Additionally, the absence of carotenoids does not necessarily lead to altered gene expression since mutants unable to synthesize carotenoids express normal levels of LHCP mRNA when they are grown in dim white light or in far-red light (Batschauer et al. 1986). The experiments described above suggest that a plastid derived factor(s), that is produced by or is sensitive to photooxidation, is necessary for the proper expression of nuclear genes that encode plastid proteins.

EXPERIMENTAL METHODS

Pea seeds (*Pisum sativa*, var. Little Marvel) were imbibed overnight in distilled water or in distilled water containing

0.1 mM norflurazon (Sandoz 9789; 4-chloro-5-(methylamino)-2(α,α,α-trifluoro-*m*-tolyl)-3(2H)-pyridazinone). The seeds were then planted in small trays in a 50:50 mixture of vermiculite and perlite. Norflurazon-treated and non-treated seeds were then exposed to complete darkness, dim light (approximately 0.01 microeinsteins per square meter per second), or bright light (approximately 100 microeinsteins per square meter per second) and grown at room temperature. The potting mixture was kept moist with either distilled water or 0.1 mM norflurazon, in accordance with the treatment the seeds initially received.

Total RNA was extracted from the leaves of 12 day old seedlings (Gantt and Key 1983), its concentration determined spectrophotometrically, and 10 micrograms was electrophoresed through 1.5% agarose/formaldehyde gels (Maniatis et al. 1982). The RNA was then transferred to nitrocellulose membranes and the levels of various mRNAs determined by hybridization. ^{32}P-labeled cDNA probes were synthesized as described by Feinberg and Vogelstein (1983).

RESULTS

After twelve days of growth, norflurazon-treated and non-treated seedlings grown in the dark were virtually indistinguishable as were treated and non-treated seedlings grown in dim light. However, herbicide-treated seedlings grown in bright light were completely bleached, except for some instances in which there was a small rim of light green color on the edge of the leaves, while the non-treated, light grown seedlings were fully green.

Figure 1 shows the results obtained when mRNAs encoding two cytosolic proteins were examined. The levels of mRNAs coding for cytosolic ribosomal proteins S11 (Gantt and Thompson, 1990) and L17 (Gantt and Key, 1985) were insignificantly affected by

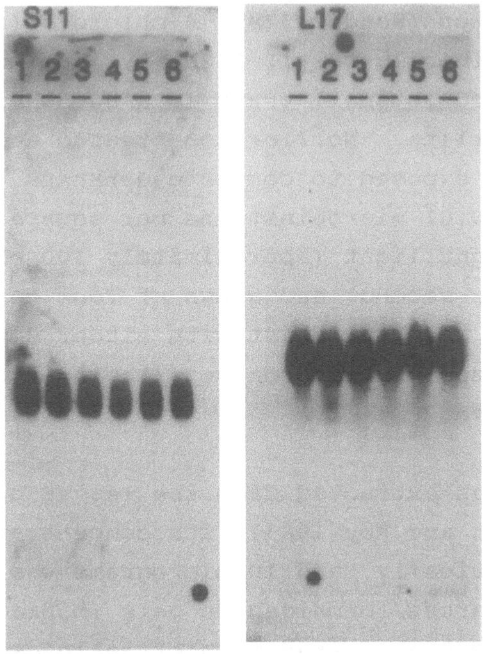

Figure 1. Northern blots of RNA extracted from control (lanes 1, 3, and 5) or norflurazon-treated seedlings (lanes 2, 4, and 6) grown in the absence of light (lanes 1 and 2), in dim light (lanes 3 and 4), or in bright light (lanes 5 and 6). The hybridizing bands correspond to mRNAs coding for cytosolic ribosomal proteins S11 and L17.

light or norflurazon treatment. The same was true for histone H1 mRNA levels (Gantt and Key, 1987; data not shown).

When the mRNA levels of nuclear-encoded chloroplast ribosomal proteins CS17 (Gantt and Thompson 1990) and CL9 (Gantt, 1988) were examined, little difference was observed between norflurazon-treated and non-treated plants grown in the dark or grown in the presence of dim light (figure 2). However, a striking difference was found in the levels of these transcripts when RNAs extracted from herbicide-treated and non-treated seedlings grown in bright light were examined. RNA dot-blot experiments showed that under conditions of bright

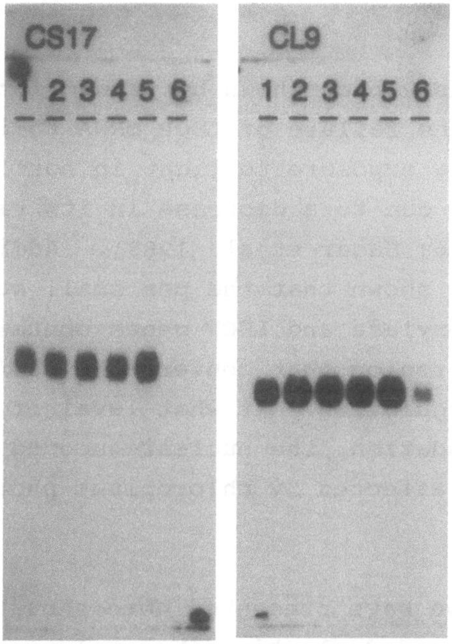

Figure 2. Northern blots of RNA extracted from control (lanes 1, 3, and 5) or norflurazon-treated seedlings (lanes 2, 4, and 6) grown in the absence of light (lanes 1 and 2), in dim light (lanes 3 and 4), or in bright light (lanes 5 and 6). The hybridizing bands correspond to mRNAs coding for nuclear-encoded plastid ribosomal proteins CS17 and CL9.

light, there was about a 10-fold decrease in hybridizing counts in the RNA extracted from the herbicide-treated plants compared to the non-treated seedlings (data not shown). Similar results were observed for five other nuclear-encoded ribosomal protein mRNAs that we were able to examine (data not shown), suggesting that perhaps all of the plastid ribosomal protein genes that are located in the nucleus are subject to this type of regulation.

DISCUSSION

It has previously been shown by transcription run-on experiments that the failure of LHCP mRNA to increase in abundance following exposure to light in norflurazon-treated maize seedlings is due to a decrease in its rate of synthesis (Taylor et al. 1986; Sagar et al. 1988). Additionally, Simpson et al. (1986) have shown that the pea small subunit of ribulose bisphosphate carboxylase and LHCP genes contain a region in their 5'-flanking region that confers the norflurazon-sensitive transcriptional regulation. At what level of regulation, synthesis or degradation, the nuclear-encoded plastid ribosomal protein mRNAs are affected by chloroplast photooxidation is not known.

Results that we have presented here show that the expression of nuclear-encoded plastid ribosomal protein genes is altered by photooxidation resulting from the lack of carotenoids. Whether the effects on the expression of these genes by norflurazon results directly from the photooxidation of a plastid derived regulatory factor or reflects a regulatory response to the loss of plastid ribosomes is not known. Further experiments are needed to understand the involvement of the chloroplast in determining the activity of the nuclear-encoded plastid ribosomal protein genes.

REFERENCES

Anderson EC, Robertson DS (1960) Role of carotenoids in protecting chlorophyll from photodestruction. Plant Physiol 35:531-534

Batschauer A, Mosinger E, Kreuz K, Dorr I, Apel K (1986) The implication of a plastid-derived factor in the transcriptional control of nuclear genes encoding the light-harvesting chlorophyll a/b protein. Eur J Biochem 154:625-634

Burgess DG, Taylor WC (1987) Chloroplast photooxidation affects the accumulation of cytosolic mRNAs encoding chloroplast proteins in maize. Planta 170:520-527

Feinberg AP, Vogelstein B (1983) A technique for radiolabeling DNA restriction endonuclease fragments to high specific activity. Anal Biochem 132:6-13

Gantt JS, Key JL (1983) Auxin-induced changes in the level of translatable ribosomal protein messenger ribonucleic acids in soybean hypocotyl. Biochemistry 22:4131-4139

Gantt JS, Key JL (1985) Coordinate expression of ribosomal protein mRNAs following auxin treatment of soybean hypocotyls. J Biol Chem 260:6175-6181

Gantt JS, Key JL (1986) Isolation of nuclear encoded plastid ribosomal protein cDNAs. Mol Gen Genet 202:186-193

Gantt JS, Key JL (1987) Molecular cloning of a pea H1 histone cDNA. Eur J Biochem 166:119-125

Gantt JS (1988) Nucleotide sequences of cDNAs encoding four complete nuclear-encoded plastid ribosomal proteins. Curr Genet 14:519-528

Gantt JS, Thompson MD (1990) Plant cytosolic ribosomal protein S11 and chloroplast ribosomal protein CS17: their primary structures and evolutionary relationships. J Biol Chem 265:2763-2767

Harpster MH, Mayfield SP, Taylor WC (1984) Effects of pigment-deficient mutants on the accumulation of photosynthetic proteins in maize. Plant Mol Biol 3:59-71

Mayfield SP, Taylor WC (1984) Carotenoid-deficient maize seedlings fail to accumulate light-harvesting chlorophyll a/b binding protein (LHCP) mRNA. Eur J Biochem 144:79-84

Maniatis T, Fritsch EF, Sambrook J (1982) Molecular cloning: a laboratory manual. Cold Spring Harbor Laboratory Cold Spring Harbor New York

Reiss T, Bergfeld R, Link G, Thein W, Mohr H (1983) Photooxidative destruction of chloroplasts and its consequences on cytosolic enzyme levels and plant development. Planta 159:518-528

Sagar AD, Horwitz BA, Elliott RC, Thompson WF, Briggs WR (1988) Light effects on several chloroplast components in norflurazon-treated pea seedlings. Plant Physiol 88:340-347

Simpson J, Van Montagu M, Herrera-Estrella L (1986) Photosynthesis-associated gene families: differences in response to tissue-specific and environmental factors. Science 233:34-38

Taylor WC, Burgess DG, Mayfield SP (1986) The use of carotenoid deficiencies to study nuclear-chloroplast regulatory interactions. Randall DD (ed) Curr Topics Plant Biochem Physiol 5:117-127

HEAT-INDUCED RIBOSOME-DEFICIENCY OF PLASTIDS - MECHANISM AND APPLICATIONS

J. Feierabend and Th. Berberich
Botanisches Institut
J. W. Goethe-Universität
P.O. Box 11 19 32
D-6000 Frankfurt am Main
Federal Republic of Germany

1. Properties and occurrence of a heat-sensitivity of 70S ribosome formation

In various higher plants, particularly in leaves of several cereal seedlings, the formation of the 70S ribosomes of the chloroplasts was observed by us to be heat-sensitive (Feierabend and Mikus, 1977). When seedlings are continuously exposed to a slightly elevated temperature in the range between 32°-34°C the new formation of 70S ribosomes is prevented in developing leaves. Because general leaf growth of cereal seedlings is only slightly retarded, 70S ribosome-deficient leaves or leaf sections are produced under non-permissive temperature conditions. Most investigations have been performed with leaves of rye seedlings. In rye leaves the accumulation of 70S ribosomes and of the plastidic rRNAs is prevented at 32°C both in light and in darkness. (Feierabend et al., 1969; Feierabend and Schrader-Reichhardt, 1976; Schäfers and Feierabend, 1979). Unlike 70S ribosome-deficiencies induced in the absence of carotene formation or under the action of chlorosis-inducing herbicides (Feierabend and Schubert, 1978; Feierabend et al., 1982) photooxidative events are not involved in the heat-dependent elimination of plastid ribosomes. An appropriate non-permissive temperature appeared to affect 70S ribosomes quite specifically. The structural differentiation or biochemical functions of other cell compartments or organelles, such as mitochondria, peroxisomes, or 80S ribosomes were not markedly impaired (Feierabend and Schrader-Reichhardt, 1976; Schäfers and Feier-

NATO ASI Series, Vol. H 55
The Translational Apparatus of Photosynthetic Organelles
Edited by R. Mache et al.

abend, 1976). However, chloroplast differentiation was defective, as can be expected from the complete absence of translation products of 70S ribosomes. Ribosome-deficient plastids remained small and proplastid-like (Feierabend et al., 1969; Schäfers and Feierabend,1976; Feierabend, 1979). While the plastid envelope had a quite normal appearance, was semipermeable and contained, for instance, functional dicarboxylate translocator activities (Höinghaus and Feierabend, 1985), the internal membrane system remained quite rudimentary, consisting only of some tubular and vesicular material (Schäfers and Feierabend, 1976; Feierabend et al., 1980). In accord with the lack of thylakoids, 70S ribosome-deficient plastids did not accumulate chlorophyll, except for minor traces. Thus, 70S ribosome-deficient leaves or leaf sections are achlorophyllous and this "heat-bleaching" can be used as indication for the absence of plastid ribosomes. In addition to rye, heat-bleaching was also observed in the leaves of several other cereal seedlings, such as wheat, oat, barley. As monitored by the decline of chlorophyll, the heat effect on plastid ribosome formation appears very sharply, within 2°C, in the range between 30°-34°C, depending on the species (Feierabend et al., 1969; Feierabend and Mikus, 1977; Feierabend, 1982). In the cereal plants leaf growth is inhibited only above 36°C, thus leaving a narrow non-permissive temperature range which allows the formation of 70S ribosome-deficient and bleached, but otherwise normal, leaves. Heat-induced bleaching and 70S ribosome-deficiency were also observed in pea at about 33°C, but in pea general leaf growth is also severely inhibited at this temperature (Feierabend and Mikus, 1977).

2. Mode of action of high temperature on 70S ribosomes

While preexisting 70S ribosomes, which are, for instance, already present before the heat treatment in the embryonic leaf cells of the dry seed, are stable and can be found in the green tip parts of 32°C-grown rye leaves, the new formation of plastid ribosomes in the newly formed leaf tissue is blocked at the non-permissive temperature (Feierabend and Schrader-Reichhardt, 1976; Feierabend, 1979). The number of plastids is not decreased by the heat treatment (Butterfass, pers. commun.). The content of plastid DNA was not diminished in 32°C-grown 70S ribosome-deficient, as compared to normal control, leaves. Restriction patterns of DNA from isolated ribosome-deficient plastids were for several restriction enzymes identical to those obtained for DNA from normal rye chloroplasts, suggesting that at least no major sequence changes had occurred (Herrmann and Feierabend, 1980). By contrast, in heat-induced bleached mutants of Euglena, which usually also lack plastid ribosomes, such as W36 ZHD or ZHB, the plastid DNA has suffered major deletions, preferentially retaining rRNA genes, and is present only in very low amounts (Heizmann et al., 1982; Hussein et al., 1982). RNA-polymerase activity of rye chloroplasts is not heat-sensitive between 30°-40°C and ribosome-deficient plastids from 32°C-grown rye leaves also contained RNA-polymerase activity and appeared to incorporate labeled uridine into RNA in vivo (Bünger and Feierabend, 1980). Accordingly, transcripts of various chloroplast genes were detected in 70S ribosome-deficient leaves by Northern hybridization with specific DNA-probes, though only in low amounts, e.g. for rbcL (Winter and Feierabend, 1990), psbA, and the common transcript for psaA, psaB and rps14 (unpublished). Though plastid rRNAs were not accumulated at the non-permissive temperature, the presence of minor amounts of processed 16S rRNA was also detected in 32°C-grown 70S ribosome-deficient leaf tissue by Northern hybridization (Fig.1). Similarly minor amounts of 16S rRNA were also detected in Norflurazon-treated 70S ribosome-deficient leaves. With respect to the evidence that RNA-polymerase genes occur on chloroplast DNA and are also expressed in the organelle (Ruf and Kössel, 1988; Little and Hallick, 1988) the detection of RNA-polymerase activity and chloroplast-specific transcripts in ribosome-deficient rye plastids demands the existence of an alternative nuclear-encoded RNA polymerase activity in the plastids. An alternative, less probable, explanation could be that the plastidic RNA-polymerase would be both extremely stable and efficient that even some highly diluted residual level of the enzyme derived through all the cell divisions during leaf growth from initial activities, possibly present in the small embryonic leaf of the dry seed, would be sufficient for the transcriptional capacity observed in totally ribosome-deficient plastids. After in vivo application of [^{3}H]-uridine to sections of 70S ribosome-deficient rye leaves at 32°C cytoplasmic rRNAs were readily labeled, while incorporation into plastidic 16S rRNA was, however, at the limit of detection. Thus, after the elimination of the 70S ribosomes current synthesis of mature plastidic rRNA

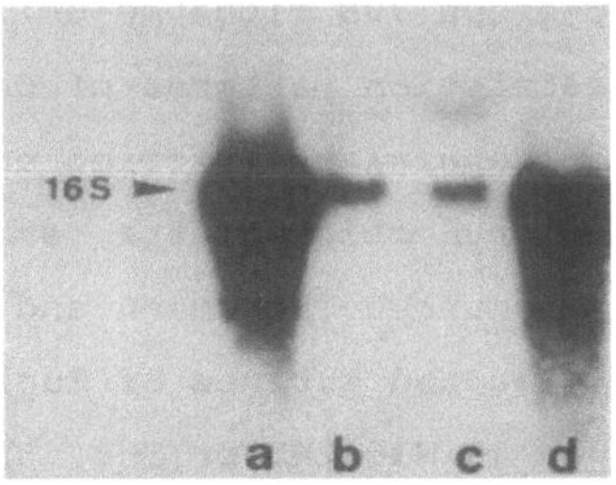

Fig. 1 Visualization of 16S rRNA by Northern hybridization of total RNA from lower halves of the leaves of 6-day-old rye seedlings using a XhoI subfragment which encodes the 16S rRNA from the PstI (7) fragment of rice chloroplast DNA as a probe (provided by Dr. Sugiura, Nagoya). a) 22°C-grown green leaves , b) 32°C-grown bleached 70S ribosome-deficient leaves, c) bleached leaves grown at 22°C in the presence of 20 μM Norflurazon, d) 22°C-grown etiolated leaves. Per lane 10 μg RNA were applied. Methods were as previously described (Winter and Feierabend, 1990).

was only extremely low or absent and also no significant accumulation of a potential rRNA precursor could be observed. Normal 70S ribosome-containing green leaves were clearly able to synthesize plastidic rRNAs also at 32°C (unpublished data).

The potential occurrence of polypeptides of plastid ribosomes was investigated in 70S ribosome-deficient tissue of rye leaves by immunoblotting with antibodies raised against purified 50S and 30S ribosomal subunits. The antibodies were able to detect about 17 of the 33 polypeptides of the 50S subunit and about 10 of the 25 polypeptides of the 30S subunit of rye chloroplasts (Feierabend et al., 1988). While the majority of these ribosomal polypeptides were not observed in soluble extracts from 70S ribosome-deficient leaves or plastids, thus confirming the absence of complete plastid ribosomes, a few ribosomal polypeptides of both the 50S and 30S subunit were clearly detected as pools of unassembled ribosomal proteins of the plastids which are of cytoplasmic origin. Though their amounts could not be exactly estimated, they were clearly more marked than the only very minor steady state levels of unassembled subunits of the chloroplast ATP synthase or of ribulosebisphosphate carboxylase in 70S ribosome-deficient leaves (Biekmann and Feierabend, 1985; Winter and Feierabend, 1990). The occurrence of unassembled ribosomal polypeptides demonstrates that coordinate controls for chloroplast ribosome formation must be less tight than the mechanisms known to synchronize ribosome biosynthesis in E. coli (Nomura et al., 1984; Lindahl and Zengel, 1986). This conception is further supported by observations that individual polypeptides of the plastid ribosomes may also form large pools of free proteins during early stages of development (Bisanz-Seyer et al., 1989) or even in mature spinach chloroplasts (Zhou and Mache, 1989). An obvious speculative possibility which needs further elucidation would be that such free ribosomal proteins have some pacemaker function for plastid ribosome assembly and are provided in advance in order to mediate the integration of other ribosomal proteins.

During long-term heat exposure the formation of most ribosomal components of the plastids appears to be suppressed or reduced to only low levels. Finally, simply the absence of ribosomal polypeptides that need to be synthesized on 70S ribosomes will restrict any further ribosome formation. However, we have so far no evidence that the synthesis of any individual ribosomal component is specifically affected by the non-permissive high temperature. At least physiological evidence is more suggestive that not the availability of any single component is limiting but that rather the assembly of the ribosomes, some processing or transport step or, alternatively, some early control step establishing the competence for plastid ribosome formation in meristematic cells, may be affected by the high temperature because short intermittent exposures of only 1 h per day can greatly restore 70S ribosome formation and greening of rye leaves also at 32°C (Feierabend, 1977). Such short time periods could conceivably allow that either free components that have each day accumulated in the newly formed tissue of the meristematic zone assemble to functional ribosomes or that some regulatory event may initiate ribosome formation. The mechanism of heat-bleaching that has also been long known for Euglena cells (Pringsheim and Pringsheim, 1952) appears to differ from that observed in higher plants. In Euglena the action of the high temperature has been ascribed to a heat-sensitivity of the plastid tRNA-polymerase (Brandt and Wiessner, 1977; Greenberg et al., 1984) in combination with a fast dilution of cell components during a stage of rapid cell division (Brandt, 1988). In heat-induced bleached Euglena mutants the chloroplast DNA also exhibits major deletions (Heizmann et al., 1981).

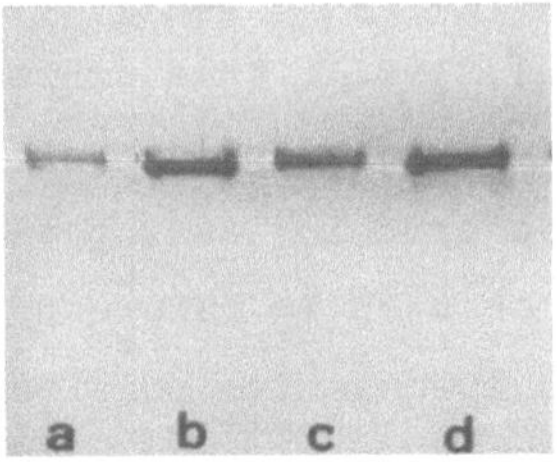

Fig. 2 Immunodetection of the HSP70 heat-shock protein after Western blotting of total extracts from the lower halves of leaves of 6-day-old rye seedlings grown in continuous light at a) 22°C, b) 32°C, c) 22°C and exposed to a heat-shock treatment at 42°C for 90 min before extraction, d) at 32°C and subjected to a heat-shock treatment as under c). The antiserum was provided by Dr. Nover, Halle/Saale, methods were previously described (Winter and Feierabend, 1990).

3. Relationships to the heat-shock response

Plants, like other organisms, exhibit the well-known and widely investigated heat-shock response which is usually induced at 40°-42°C (for review see Neumann et al., 1989) and characterized by the specific synthesis of certain heat-shock proteins, while translation of proteins for normal house-keeping functions is turned off. Representatives of both the small HSP20 and the large HSP70 heat-shock protein families have also been located in chloroplasts (Kloppstech et al., 1985; Vierling et al., 1988; Chen et al., 1990; Marshall et al., 1990). Inasmuch as heat-shock proteins may have properties of chaperonins affecting transport and assembly processes (see Marshall et al., 1990) effects on plastid ribosomes appear conceivable. During prolonged exposure to temperatures of 32° or 33°C both large heat-shock proteins HSP70 (Fig. 2) and small heat-shock proteins HSP21 (unpublished results) were induced to their maximal levels. However, in contrast to true heat-shock conditions the synthesis of the majority of normal proteins was not turned off at 32°C. Short heat-shock treatments at 40°C could not modify the 70S ribosome-deficiency observed at 32°C by any preadaptation. We were also unable to mimic 70S ribosome-deficiency, as observed at 32°C, at a permissive temperature by regular intermittent short heat-shock treatments which would also maintain heat-shock proteins at a high level. In conclusion, while heat-shock-proteins are induced during continuous growth at 32°C we have no indication for any connection to the block of 70S ribosome formation.

4. Applications of heat-bleached 70S ribosome-deficient leaves

The heat-induced deficiency of 70S ribosomes may serve as convenient experimental approach for the specific long-term elimination of plastidic translation avoiding the use of inhibitors with potential toxic side effects. Questions of the contribution of the nuclear-cytoplasmic compartment to the biosynthesis of chloroplasts and of regulatory interrelationships coordinating the biosynthesis of chloroplast components inside and outside the organelle have been addressed by the use of heat-treated 70S ribosome-deficient leaves.

A great number of chloroplast proteins or enzymes made on 80S ribosomes were detected in 70S ribosome-deficient leaves or plastids. These include enzymes involved in photosynthesis (Feierabend et al., 1969; Feierabend and Schrader-Reichhardt, 1976; Feierabend et al., 1984; Biekmann and Feierabend, 1985; Winter and Feierabend, 1990), carbohydrate metabolism (Feierabend and Gringel, 1983; Kurzok and Feierabend, 1986; Otto and Feierabend, 1989), nitrogen metabolism (Feierabend, 1986), lipid metabolism (Kosmac and Feierabend, 1985), chlorophyll (Feierabend, 1977; Batschauer et al., 1982) and pyrimidine biosynthesis (Doremus and Jagendorf, 1987), ribosomal (Feierabend et al., 1988) and membrane polypeptides (Feierabend et al., 1980), and translocators of the plastid envelope (Höinghaus and Feierabend, 1985). The quantities of most soluble photosynthetic enzymes moderately decreased in 70S ribosome-deficient leaves amounting to 30-70 % of their levels in normal leaves (Feierabend and Schrader-Reichhardt, 1976; Feierabend et al., 1984). The expression of other polypeptides of cytoplasmic origin involved in photosynthesis, particularly of those which are constituents of macromolecular complexes or membranes, was more strongly suppressed

(Feierabend et al., 1984; Biekmann and Feierabend, 1985; Winter and Feierabend, 1990). However, several isoenzymes of sugar phosphate metabolism, which are more important under non-photosynthetic conditions, were higher in bleached ribosome-deficient plastids than in chloroplasts (Kurzok and Feierabend, 1986; Otto and Feierabend, 1989). This differential behavior needs to be considered for an understanding of interactions between the nuclear and plastid genomes. From observations that the expression of the light-harvesting chlorophyll a/b protein or the small subunit of the ribulosebisphosphate carboxylase were greatly reduced after elimination of chloroplast translation, particularly after application of bleaching herbicides, like Norflurazone evoking a photodegradative loss of 70S ribosomes, it has been deduced that nuclear transcription of chloroplast protein genes is dependent on some factor to be transmitted from functional chloroplasts (Oelmüller, 1989; Taylor, 1989). From our comparative observations we would suggest that the role of such "plastid factor" signals would not so much lie in the specific induction or repression of single proteins but their function may rather be to switch between whole patterns of alternative enzyme or protein complements for the differentiation of either photosynthetic chloroplasts or non-photosynthetic achlorophyllous plastids, in accord with the needs of the different metabolic conditions of the tissues. The response might be mediated by regulatory elements, normally specifying tissue-specific expression of the genes.

As examples for studying regulatory interactions between the chloroplasts and the nucleo-cytoplasm we have, in particular, investigated the interdependance of pigment and apoprotein formation of the light-harvesting chlorophyll a/b complex and the coordinate expression of subunit polypeptides of different genetic origin of the ribulosebisphosophate carboxylase, the coupling factor of the ATP synthase CF_1, and the plastid ribosome.

There is no principal need that ribosome-deficient plastids are bleached. Though the chloroplast-encoded chlorophyll-binding reaction center polypeptides of the photosystems are missing, 70S ribosome-deficient leaves should be able to synthesize the nuclear-encoded light-harvesting chlorophyll a/b protein. However, the synthesis of the light-harvesting chlorophyll a/b protein, as estimated by in vivo [^{35}S]-methionine incorporation, is greatly reduced and already restricted by a correspondingly low level of mRNA which is only between 10-15 % of the amount in normal green rye leaves (unpublished results). Due to a slow degradation a steady state level of only about 1 % of the light-harvesting chlorophyll a/b protein found in green leaves was detected in 32°-grown 70S ribosome-deficient leaf tissue (Feierabend et al., 1984). Though traces of chlorophyll were present including both chlorophylls a and b, its quantity accounted for less than 1%, although the enzymes of chlorophyll synthesis appear to be nuclear-encoded and present in 70S ribosome-deficient leaves (Feierabend, 1977), including the protochlorophyllide reductase (Batschauer et al., 1982) and the phytylating chlorophyll synthetase (Blank-Huber, 1986). The synthesis of δ-aminolevulinic acid appears to be suppressed by a feedback inhibition. After application of δ-aminolevulinic acid, however, major amounts of protochlorophyll(ide) are accumulated in darkness but virtually not phototransformed upon illumination, and the appearance of the light-harvesting chlorophyll a/b protein is not markedly stimulated. Extensive studies, comparing translation and mRNA levels of the light-harvesting chlorophyll a/b protein after applications of δ-aminolevulinic acid or various inhibitors of chlorophyll biosynthesis, suggest that the synthesis of the light-harvesting chlorophyll a/b protein was at least under short-term conditions, not dependent on concomitant chlorophyll synthesis and that precursors or intermediates of chlorophyll synthesis did not play any relevant role as signals for its induction or suppression (Bothe and Feierabend, unpublished results). However, the synthesis of the light-harvesting chlorophyll a/b protein and presumably also the assembly of the pigment-protein-complex appeared to be strongly coupled to the availability of 70S ribosome translation products. In green rye leaves incorporation into the light-harvesting chlorophyll a/b protein was strongly decreased by fast translational controls before changes of the mRNA level were detectable when translation in the chloroplasts was blocked, e.g. by chloramphenicol, whereas inhibitors of chlorophyll synthesis had no comparable effects. We conclude that the formation of the light-harvesting chlorophyll a/b protein can only follow the appearance of the core complexes of the photosystems and that the absence of 70S ribosome translation products is responsible for its suppression in 32°-grown leaves and consequently also for the bleaching.

The system of 32°C-grown 70S ribosome-deficient rye leaves has also been applied for the analysis of regulatory interactions that serve to establish precise subunit stoichiometries during the biogenesis of multimeric proteins, such as the ribulosebisphosphate carboxylase and the coupling factor of the ATP synthase CF_1. In rye leaves multiple controls appear to contribute to a coordinate expression of ribulosebisphosphate carboxylase subunits. The synthesis of the large subunit in the chloroplast and of the small subunit in the cytoplasm is coordinated by fast translational controls and by an attuned supply of mRNAs. These mechanisms by which the rates of synthesis of the two subunits are always greatly adapted to each other, require an exchange of regulatory signals between the compartments that must be transmitted in both directions (Winter and Feierabend, 1990). However, controls acting at the transcript and translational level are somewhat leaky and allow a low rate of synthesis of each subunit in the absence of the other. Thus, in 70S ribosome-deficient rye leaves free small subunit was synthesized in the absence of large subunit. (Feierabend and Wildner, 1978). However, the rate of its synthesis was restricted to only about 5 %, relative to normal leaves, by a similarly low level of its mRNA (Winter and Feierabend, 1990). Obviously due to the action of strong mechanisms providing an attuned mRNA supply for both subunits, the small subunit mRNA in 70S ribosome-deficient leaves was accompanied by a low level of large subunit mRNA which also still responded to changes of the small subunit mRNA. When the synthesis of the small subunit in 70S ribosome-deficient leaves was blocked by an inhibitor of 80S ribosomal translation not only the mRNA for the small subunit but also that for the large subunit further declined quite markedly. This observation suggests that a suppression of the large subunit mRNA level occurring under conditions when the small subunit supply is restricted cannot be exerted by feedback effects of accumulating free large subunits, since these do not occur in ribosome-deficient plastids (Winter and Feierabend, 1990).

In 70S ribosome-deficient leaves not only the synthesis of unassembled small subunits of ribulosebisphosphate carboxylase but also of γ- and δ-subunits of the ATP synthase coupling factor CF_1 has been observed (Biekmann and Feierabend, 1985). However, these free subunit polypeptides were not accumulated, but rapidly degraded, whereas they exhibit high stability when integrated into their holoproteins in normal leaves (Biekmann and Feierabend, 1985; Winter and Feierabend, 1990). The rapid proteolysis of unassembled nuclear-encoded subunit polypeptides appears to provide a final fine tuning mechanism to establish a precise subunit stoichiometry of such multimeric proteins. The use of in vivo labeled unassembled subunits made in 32°-grown 70S ribosome-deficient rye leaves enabled us to establish an in vitro assay system for an initial characterization of the proteolytic activities involved. Properties of the proteases acting on free small subunits or on the holoprotein of ribulosebisphosphate carboxylase and on CF_1-δ were clearly different. Thiol-proteases were involved in the degradation of the small subunit of ribulosebisphosphate carboxylase whereas serine-proteases appeared to act on CF_1-δ (Feierabend et al., 1990).

5. Acknowledgements

Support of our work by the Deutsche Forschungsgemeinschaft and by the Fonds der Chemischen Industrie is greatly appreciated. We thank Prof. Lutz Nover, Halle/Saale, for his cooperation and the DAAD, Bonn, for a travel grant (to Th. B.).

6. References

Batschauer A, Santel HJ, Apel K (1982) The presence and synthesis of the NADPH-protochlorophyllide oxidoreductase in barley leaves with a high temperature-induced deficiency of plastid ribosomes. Planta 154:459-464

Biekmann S, Feierabend J (1985) Synthesis and degradation of unassembled polypeptides of the coupling factor of photophosphorylation CF_1 in 70S ribosome-deficient rye leaves. Eur J Biochem 152:529-535

Bisanz-Seyer C, Li YF, Peyer P, Mache R (1989) The components of the plastid ribosome are not accumulated synchronously during the early development of spinach plants. Plant Mol Biol 12:201-211

Blank-Huber M (1986) Untersuchungen zur Chlorophyll-Biosynthese. Stabilisierung und Eigenschaften der Chlorophyll-Synthetase. Doctoral thesis University München

Brandt P (1988) Strain and stage-specific high-temperature treatment of Euglena gracilis causes permanent apochlorosis. J Plant Physiol 133:281-287

Bünger W, Feierabend J (1980) Capacity for RNA synthesis in 70S ribosome-deficient plastids of heat-bleached rye leaves Planta 149:163-169

Chen Q, Lauzon LM, DeRocher AE, Vierling E (1990) Accumulation, stability, and localization of a major chloroplast heat-shock protein. J Cell Biol 110:1873-1883

Doremus HD, Jagendorf AT (1987) Site of synthesis of the enzymes of the pyrimidine biosynthetic pathway in oat (Avena sativa L.)leaves. Plant Physiol 83:657-658

Feierabend J (1977) Capacity for chlorophyll synthesis in heat-bleached 70S ribosome-deficient rye leaves. Planta 135:83-88

Feierabend J (1979) Role of cytoplasmic protein synthesis and its coordination with the plastidic protein synthesis in the biogenesis of chloroplasts. Ber Deutsch Bot Ges 92:553-574

Feierabend J (1982) Inhibition of chloroplast ribosome formation by heat in higher plants. In: Methods in chloroplast molecular biology, pp. 671-680, Edelman M, Hallick RB, Chua NH (eds) Elsevier, Amsterdam

Feierabend J (1986) Investigation of the site of synthesis of chloroplastic enzymes of nitrogen metabolism by the use of heat-treated 70S ribosome-deficient rye leaves. Physiol Plant 67:145-150

Feierabend J, Gringel G (1983) Plant transketolase: subcellular distribution, search for multiple forms, site of synthesis. Z Pflanzenphysiol 110:247-258

Feierabend J, Mikus M (1977) Occurrence of a high temperature sensitivity of chloroplast ribosome formation in several higher plants. Plant Physiol 59:863-867

Feierabend J, Schrader Reichhardt U (1976) Biochemical differentiation of plastids and other organelles in rye leaves with a high-temperature-induced deficiency of plastid ribosomes. Planta 129:133-145

Feierabend J, Schubert B (1978) Comparative investigation of the action of several chlorosis-inducing herbicides on the biogenesis of chloroplasts and leaf microbodies. Plant Physiol 61:1017-1022

Feierabend J, Wildner G (1978) Formation of the small subunit in the absence of the large subunit of ribulose 1,5-bisphosphate carboxylase in 70S ribosome-deficient rye leaves. Arch Biochem Biophys 186:283-291

Feierabend J, Berger Ch, Meyer A (1969) Spezifische Störung von Entwicklung und Enzymbildung der Plastiden höherer Pflanzen durch hohe Wachstumstemperaturen. Z Naturforschg. 24b:1641-1647

Feierabend J, Meschede D, Vogel KD (1980) Comparison of the polypeptide compositions of the internal membranes of chloroplasts, etioplasts and ribosome-deficient heat-bleached plastids from rye leaves. Z Pflanzenphysiol 98:61-78

Feierabend J, Winkelhüsener T, Kemmerich P, Schulz U (1982) Mechanism of bleaching in leaves treated with chlorosis-inducing herbicides. Z Naturforsch 37c:898-907

Feierabend J, Biekmann S, Höinghaus R, Kosmac U (1984) Investigation of chloroplast membrane formation and its control with the aid of heat-bleached 70S ribosome-deficient leaves. Isr J Bot 33:93-106

Feierabend J, Schlüter W, Tebartz K (1988) Unassembled polypeptides of the plastidic ribosomes in heat-treated 70S-ribosome-deficient rye leaves. Planta 174:542-550

Feierabend J, Bergmann C, Otto S (1990) Comparison of properties of the proteolytic degradation of unassembled nuclear-encoded subunits of ribulose-1,5-bisphosphate carboxylase and of the coupling factor of photophosphorylation CF_1. Bot Acta 103: 174-182

Greenberg BM, Narita JO, de Luca-Flaherty C, Gruissem W, Rushlow KA, Hallick RB (1984) Evidence for two RNA polymerase activities in Euglena gracilis chloroplasts. J Biol Chem 259: 14880-14887

Heizmann P, Doly J, Hussein Y, Nicolas P, Nigon V, Bernardi G (1981) The chloroplast genome of bleached mutants of Euglena gracilis. Biochim Biophys Acta 653:412-415

Herrmann RG, Feierabend J (1980) The presence of DNA in ribosome-deficient plastids of heat-bleached rye leaves. Eur J Biochem 104:603-609

Höinghaus R, Feierabend J (1985) Origin and developmental changes of envelope proteins and translocator activities from plastids of Secale

cereale L. Planta 166:452-465
Hussein Y, Heizmann P, Nicolas P, Nigon V (1982) Quantitative estimations of chloroplast DNA in bleached mutants of Euglena gracilis. Curr Genet 6:111-117
Kloppstech K, Meyer G, Schuster G, Ohad I (1985) Synthesis,transport and localization of a nuclear coded 22-kd heat-shock protein in the chloroplast membranes of peas and Chlamydomonas reinhardii. EMBO J 4:1901-1909
Kosmac U, Feierabend J (1985) Control of plastidic glycolipid synthesis and its relation to chlorophyll formation. Plant Physiol 79:646-652
Kurzok HG, Feierabend J (1986) Comparison of the development and site of synthesis of a cytosolic and a chloroplast isoenzyme of triosephosphate isomerase in rye leaves. J Plant Physiol 126: 207-212
Lindahl L, Zengel JM (1986) Ribosomal genes in Escherichia coli. Annu Rev Genet 20:297-326
Little MC, Hallick RB (1988) Chloroplast rpoA, rpoB, and rpoC genes specify at least three components of a chloroplast DNA-dependent RNA polymerase active in tRNA and mRNA transcription. J Biol Chem 263:14302-14307
Marshall S, DeRocher AE, Keegstra K, Vierling E (1990) Identification of heat shock protein hsp70 homologues in chloroplasts. Proc Natl Acad Sci USA 87:374-378
Neumann D, Nover L, Parthier B, Rieger R, Scharf KD, Wollgiehn R, zur Nieden U (1989) Heat shock and other stress response systems of plants. Biol Zentbl 108:1-156
Nomura M, Gourse R, Boughman G (1984) Regulation of the synthesis of ribosomes and ribosomal components. Annu Rev Biochem 53:75-117
Oelmüller R (1989) Photooxidative destruction of chloroplasts and its effect on nuclear gene expression and extraplastidic enzyme levels. Photochem Photobiol 49:229-239
Otto S, Feierabend J (1989) Enzymes of starch and sugar phosphate metabolism in achlorophyllous ribosome-deficient plastids from high-temperature-grown rye leaves. Physiol Plant 76:65-73
Pringsheim EG, Pringsheim O (1952) Experimental elimination of chromatophores and eyespot in Euglena gracilis. New Phytol 51:65-68
Ruf M, Kössel H (1988) Structure and expression of the gene coding for the a-subunit of DNA-dependent RNA polymerase from the chloroplast genome of Zea mays. Nucl Acid Res 16:5741-5754
Schäfers HA, Feierabend J (1976) Ultrastructural differentiation of plastids and other organelles in rye leaves with a high- temperature-induced deficiency of plastid ribosomes. Cytobiologie (Eur J Cell Biol) 14:75-90
Taylor WC (1989) Regulatory interactions between nuclear and plastid genomes. Annu Rev Plant Physiol 40:211-233
Vierling E, Nagao RT, DeRocher AE, Harris LM (1988) A heat shock protein localized to chloroplasts is a member of a eukaryotic superfamily of heat shock proteins. EMBO J 7:575-582
Winter U, Feierabend J (1990) Multiple coordinate controls contribute to a balanced expression of ribulose-1,5-bisphosphate carboxylase/oxygenase subunits in rye leaves. Eur J Biochem 187:445-453
Zhou DX, Mache R (1989) Presence in the stroma of chloroplasts of a large pool of a ribosomal protein not structurally related to any Escherichia coli ribosomal protein. Mol Gen Genet 219:204-208

REGULATION OF PROTEIN SYNTHESIS IN CHLOROPLASTS OF CHLAMYDOMONAS REINHARDII

A. Boschetti, R. Blättler and E. Breidenbach,
Institut für Biochemie, Universität Bern,
Freiestrasse 3
CH-3012 Bern
Switzerland

The chloroplast genome contains about 100 genes for proteins. The biogenesis of the chloroplast, however, requires at least 5 to 10 times more protein species. Most of these peptides are encoded in the nuclear genome, synthesized as precursors in the cytoplasm, and transported, while processed, into the chloroplast. Many multimeric protein complexes of the chloroplast, e.g. Rubisco[1], PS II-core complexes, ribosomes, are built up stoichiometrically by peptide-subunits from both origins, the cytoplasm and the chloroplast. Since the pool of free subunits in the plastid is usually rather low, a cooperation of the nuclear and chloroplast genetic systems is necessary to keep the synthesis of partner proteins in step and for an efficient synthesis of the holocomplexes. This becomes especially obvious during chloroplast biogenesis or when the enzyme equipment of the plastid has to be adapted to new environmental or developmental conditions. The mechanism of this regulation between the nucleo-cytoplasm and the chloroplast is still unknown.

In the last ten years the concept grew up that posttranscriptional regulation of protein synthesis is of major importance

[1]Abbreviations:
Rubisco: ribulose-1,5-bisphosphate carboxylase; LS: large subunit of Rubisco; PS II: photosystem II; D1-protein: herbicide-binding 32 kDa membrane protein or Q_B-protein of PS II; EF-Tu: translational elongation factor Tu.

NATO ASI Series, Vol. H 55
The Translational Apparatus of Photosynthetic Organelles
Edited by R. Mache et al.

within the chloroplast itself (for review see Jenkins et al. 1983, Link 1988, Gruissem 1989, Taylor 1989, Boschetti et al. 1990). The bulk of evidence for translational or posttranscriptional regulation of chloroplast gene expression is based on studies during differentiation of plastids. During light-induced transformation of etioplasts to chloroplasts the rate of synthesis of individual proteins in the plastid increased much more than the amount of their mRNA. However, posttranscriptional regulation of protein expression in plastids is not restricted to differentiation events. In green algae new chloroplasts are formed not by differentiation of proplastids or etioplasts, but by division of mature chloroplasts. Therefore, in synchronized cultures of algae regulation of chloroplast protein synthesis can be observed under rather physiological conditions during the normal cell cycle in fully differentiated chloroplasts.

Working with synchronized cultures of Chlamydomonas reinhardii we are using this last approach to adress the following questions: (a) How is the expression of specific proteins turned on and off during the cell cycle? (b) Has the binding of ribosomes and polysomes to thylakoids something to do with such a regulation? (c) What kind of signals are responsible for such a regulation? Do the signals (or regulatory factors) come from the cytoplasm or from the chloroplast itself?

EXPERIMENTAL

The type of experiments to measure the rates of synthesis of individual chloroplast proteins during the cell cycle is exemplified in Fig.1. In synchronized cultures of Chlamydomonas reinhardii chloroplast protein synthesis is most prominent in the light (Fig.1c). The rates of synthesis of three proteins were measured during the cell cycle, namely of (i) the membrane protein D1, (ii) the soluble protein LS, and (iii) the soluble protein EF-Tu, which is not directly involved in photosynthesis. It should be mentioned that EF-Tu in Chlamydomonas reinhardii is encoded in the chloroplast DNA.

The identification of these proteins on polyacrylamide gels was described previously (Leu et al. 1984; Breidenbach et al. 1990). EF-Tu is serologically detectable during the whole cell cycle (Fig.1e), but its synthesis is modulated as shown by the incorporation of radioactive tracers (Fig.1c). To quantify

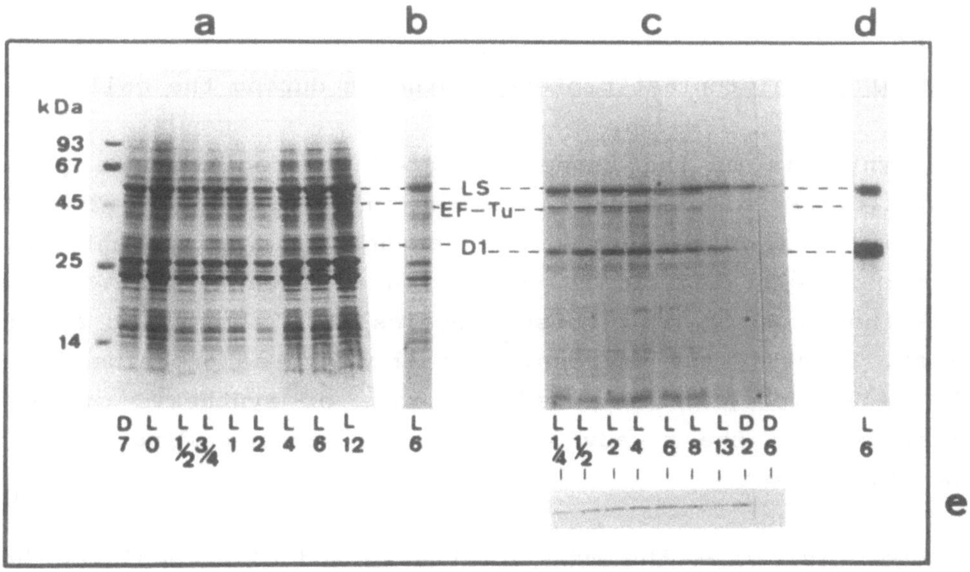

Figure 1: (a): Whole cells were pulse labeled for 60 min. with [^{35}S]sulfate in the presence of cycloheximide at different hours of the cell cycle (L: light period; D: dark period; the numbers indicate the hour of the respective period). Cycloheximide inhibits protein synthesis on the cytoplasmic 80S ribosomes, but not on the chloroplast 70S ribosomes. Coomassie Blue staining of electrophoretically separated cell proteins on SDS-polyacrylamide gel. (b): Isolated chloroplasts from cells harvested at the 6th hour in the light were pulse labeled for 20 min. with [^{35}S]methionine. Coomassie Blue staining. (c): Autoradiogram of (a). (d): Autoradiogram of (b). (e): Immunostaining of Western blot of (c) with antiserum against E.coli EF-Tu. The band identified as EF-Tu, and migrating slightly above protein P6, is present during the whole cell cycle.

this modulation, the proteins were cut out from the polyacrylamide gel and the incorporated radioactivity was measured.

The amounts of total chloroplast RNA and of specific mRNA were measured in isolated pure chloroplasts. The mRNA content was determined by hybridizing labeled gene probes with Northern blots of RNA extracted from isolated chloroplasts and measuring the radioactivity (Breidenbach et al. 1988, 1990).

RESULTS

Control of chloroplast protein synthesis during the cell cycle

The synthesis of the three mentioned chloroplast-made proteins, i.e. EF-Tu, D1 and LS, occured in the light (Fig.2a). The maximal rate of synthesis was reached at a different time, pointing to an individual regulation of each protein. In the dark (hours D0 to D10) these proteins were almost not synthesized. To study the expression of these three proteins during the cell cycle, we compared the rate of synthesis of these proteins with the steady state of the corresponding mRNAs in the chloroplast.

As shown in Fig.2b the mRNA for the LS and D1 accumulated during the light period and reached a four times higher level at the end of it. The cells in our synchronized cultures divide in the average into four daughter cells during the first half of the dark period. Obviously, the mRNA content for LS and D1 is distributed among the four daughter cells to reach the relative mRNA content per cell of about 1 at the beginning of the next dark period. A different pattern is reflected in the mRNA content for EF-Tu. Following the increase at the beginning of the light period its abundance declined to a very low level at the end of it. In order to explain the substantial amount of mRNA in the doughter cells at the beginning of the next light period a transcriptional activity somewhen in the dark period has to be presumed.

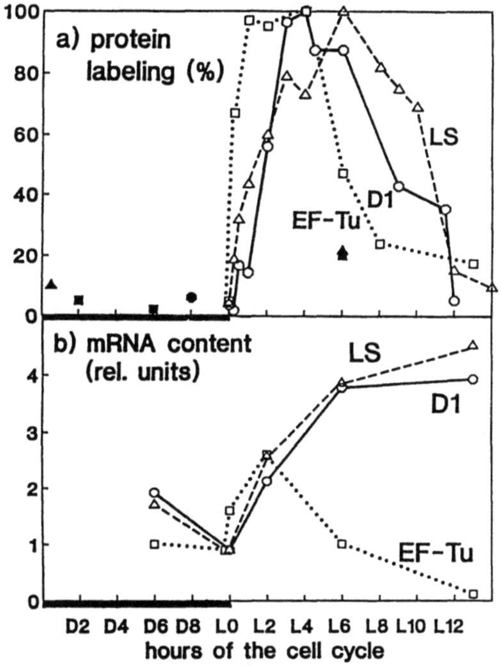

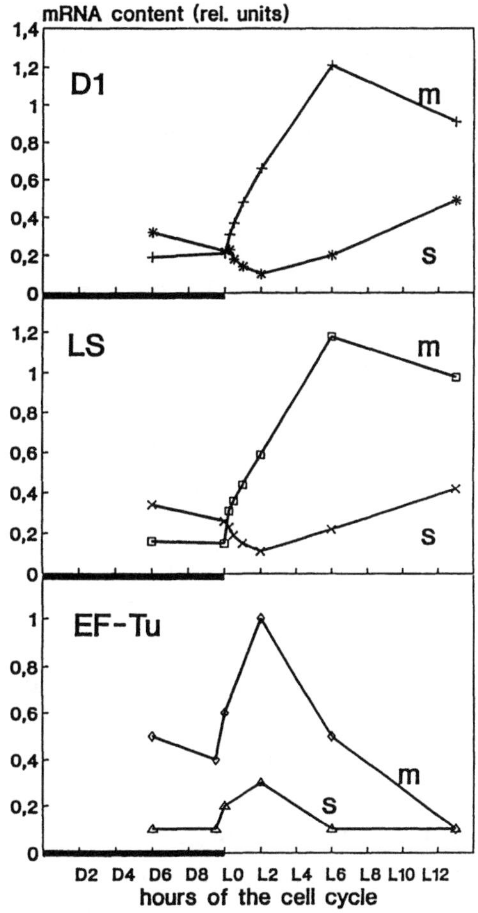

Figure 2: a) Rate of protein synthesis of D1 (○), LS (△), and EF-Tu (□) during the cell cycle. Labeling in the light: ○,△,□; in the dark: ●, ▲, ■. For each protein the highest incorporation was taken as 100%. b) mRNA-content in the chloroplast during the cell cycle.

Figure 3: Distribution of the mRNA for D1, LS and EF-Tu between membrane bound polysomes (m) and free polysomes from the stroma (s).

Comparing the mRNA levels (Fig.2b) with the rates of protein synthesis (Fig.2a) we conclude:

(1) The synthesis of LS and D1 is not directly driven by the

abundance of mRNA. At the end of the light period, although the mRNA content in chloroplasts is high, there is almost no protein synthesis. Therefore, the expression of D1 and LS might be regulated by translational activity.

(2) The rate of EF-Tu synthesis paralleled the content of the mRNA during the light period. Therefore the expression of EF-Tu might be regulated by mRNA abundance during the light period but presumably not in the dark.

(3) A complementary conclusion comes from experiments on protein synthesis in isolated chloroplasts. Although mRNA for the above mentioned proteins exist during the dark period in the cell, protein synthesis in chloroplasts isolated from cells during the dark period can not be stimulated by incubation in the light, whereas in chloroplasts from illuminated cells the proteins are readily made. Therefore one has to look also for some cell cycle dependent regulation mechanisms.

Localization of ribosomes and mRNA in the chloroplast

In green cells light stimulates not only protein synthesis but also the binding of chloroplast ribosomes and polysomes to thylakoids, whereas in the dark the amount of membrane bound ribosomes and polysomes is low. During the dark-to-light transition of sychronized Chlamydomonas cultures the amount of RNA was found to increase sharply in the thylakoid and to decrease in the stroma fractions. In the dark, half of the RNA was membrane bound, whereas in the light most of the RNA was immediately bound to thylakoids. At the end of the light period, when protein synthesis was low, we found an equal distribution between thylakoids and stroma.

What happened with the individual mRNAs during the cell cycle? Fig.3 shows that about 80% of the mRNA for the LS and D1 was membrane bound when protein synthesis was maximal. Also the mRNA for EF-Tu was accumulated in thylakoid bound polysomes. This distribution paralleled once again the rate of EF-Tu synthesis. The finding that the mRNAs for the soluble proteins were also enriched in thylakoid bound polysomes contradicts the hypothesis that only membrane proteins are translated on thylakoid bound polysomes. Moreover, we believe that polysome bindig to thylakoids plays some important, but still unknown role for translation in the chloroplast. One could ask whether initiation and/or elongation are influenced by a membrane bound factor.

Cytoplasmic influence on chloroplast translation in vitro

Genetic analysis of some phenotypic PSII-mutants revealed that certain nuclear mutations may inhibit the translation of selected mRNA in the chloroplast (Jensen et al. 1986; Kuchka et al. 1988; Rochaix et al. 1989; Gamble and Mullet 1989). To study such cytoplasmic influence on chloroplast translation we were trying to modify the in vitro protein synthesis of isolated chloroplasts by a cytoplasmic soluble fraction. We isolated functionally active chloroplasts from cells harvested at the 6th hour in the light and we measured the light driven incorporation of [^{35}S]methionine into chloroplast-made proteins, especially into LS and D1. The product pattern on SDS polyacrylamide gels of this in organello synthesis (Fig.1b,d) was very similar to the in vivo pattern obtained in the presence of cycloheximide (Fig.1a,c).

In a further step we tried to modify such in organello protein synthesis by the addition of a soluble cytoplasmic supernatant S100. Fig.4 shows that the addition of cytoplasmic S100 derived from cells harvested at the middle of the light period (L6) stimulated, from the end of the light period or from the dark period (L9, L11, D2) inhibited protein synthesis.

Furthermore, we observed that the ratio of protein synthesis of D1 to LS is different in isolated chloroplasts (in organello) and in cycloheximide treated cells (in vivo) (Fig.5).

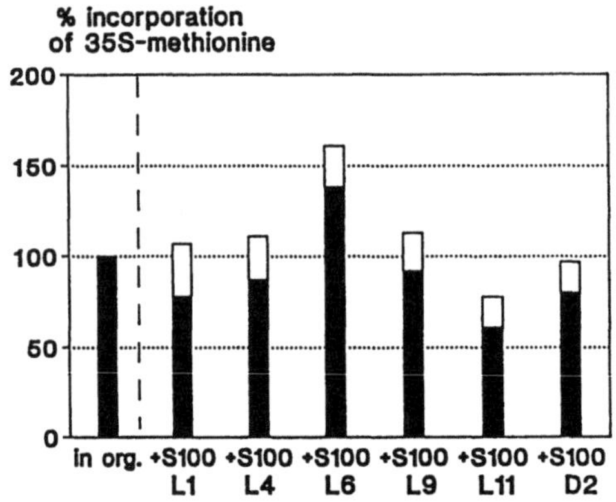

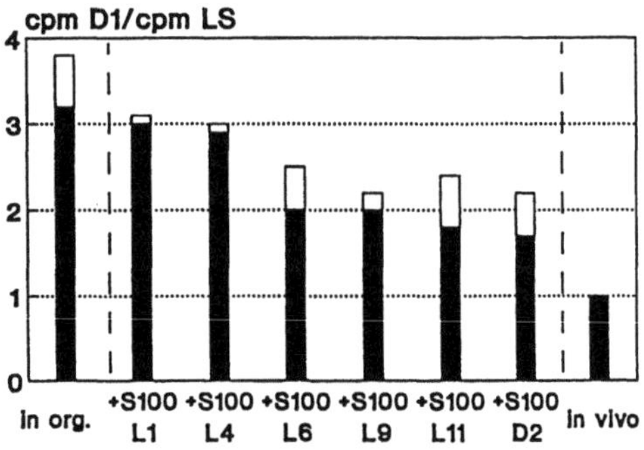

Figure 4: Modification of protein synthesis in isolated chloroplasts by addition of cytoplasmic supernatants S100 from cells harvested at the indicated hours of the cell cycle. Incorporation of [^{35}S]-methionine into chloroplasts isolated from cultures at the 6th hour of the light period was taken as 100%. To such incorporation assays S100 was added, prepared from cells harvested at different time during the cell cycle. Black bars: mean values of at least four up to 18 experiments; open bars: standard deviation.

Figure 5: Ratio of protein synthesis of D1 to LS in isolated chloroplasts modified by addition of S100. Same experiment as in Fig.4 after electrophoretic separation of the proteins.

Obviously, the addition of cytoplasmic supernatants from different times during the cell cycle not only changed the total incorporation rate, but modified the synthesis of LS and D1 individually. Supernatants from cells grown to the late light or even dark period shifted the ratio more towards the in vivo situation. At the moment we are trying to isolate and characterise these modifying factors.

CONCLUSIONS

Control of chloroplast protein synthesis during the cell cycle

(i) In synchronized green algae the synthesis of the two chloroplast-made proteins LS and D1 do not follow the variation in the mRNA content. This points to a translational control of these proteins.

(ii) EF-Tu synthesis parallels the mRNA-content during the light period, and seems therefore to be regulated by transcription or by stability of mRNA during the light period.

Thylakoids and protein synthesis

In chloroplasts the synthesis of membrane-bound as well as of soluble proteins is associated with thylakoids.

Influence of cytoplasmic factors on chloroplast translation

We have preliminary data, which suggest that one or more cytoplasmic components inhibit or stimulate the rate of synthesis of individual proteins in isolated chloroplasts. The effect of these components is cell cycle dependent.

ACKNOWLEDGEMENTS

The identification of EF-Tu was made possible by the generous gift of antisera against E.coli EF-Tu by Drs E.Vygenboom and A.Parmeggini. This work was partly suported by the Swiss National Foundation for Scientific Research (grant No.31-27747.89).

REFERENCES

Boschetti A, Breidenbach E, Blättler R (1990) Control of protein formation in chloroplasts. Plant Sci 68:131-149

Breidenbach E, Jenni E, Leu S, Boschetti A (1988) Quantification of mRNA in chloroplasts of Chlamydomonas reinhardii: Equal distribution of mRNA for a soluble and a membrane polypeptide between stroma and thylakoids. Plant Cell Physiol. 29:1-7

Breidenbach E, Leu S, Michaels A, Boschetti A (1990) Synthesis of EF-Tu and distribution of its mRNA between stroma and thylakoids during the cell cycle of Chlamydomonas reinhardii. Biochim. Biophys. Acta 1048:209-216

Gamble PE, Mullet JE (1989) Translation and stability of proteins encoded by the plastid psbA and psbB genes are regulated by a nuclear gene during light-induced chloroplast development in barley. J Biol Chem 264:7236-7243

Gruissem W (1989) Chloroplast gene expression: How plants turn their plastids on. Cell 56:161-170

Jenkins GI, Hartley MR, Bennett J (1983) Photoregulation of chloroplast development: translational, transcriptional and posttranslational control? Phil Trans R Soc Lond B303:419-431

Jensen KH, Herrin DL, Plumley FG, Schmidt GW (1986) Biogenesis of photosystem II complexes: transcriptional, translational and posttranslational regulation. J Cell Biol 103:1315-1325

Kuchka MR, Mayfield SP, Rochaix JD (1988) Nuclear mutations specifically affect the synthesis and/or degradation of the chloroplast-encoded D2 polypeptide of photosystem II in Chlamydomonas reinhardtii. EMBO J 7:319-324

Leu S, Bolli R, Mendiola-Morgenthaler L, Boschetti A (1984) In-vitro translation of different mRNA-containing fractions of Chlamydomonas chloroplasts. Planta 160:204-211

Link G (1988) Photocontrol of plastid gene expression. Cell Environ 11:329-338

Rochaix JD, Kuchka M, Mayfield S, Schirmer-Rahire M, Girard-Bassecou J, Bennoun P (1989) Nuclear and chloroplast mutations affect the synthesis or stability of the chloroplast psbC gene product in Chlamydomonas reinhardtii. EMBO J 8:1013-1021

Taylor WC (1989) Regulatory interactions between nuclear and plastid genomes. Annu Rev Plant Physiol Plant Mol Biol 40:211-233

PARTICIPANTS AND CONTRIBUTORS

Mr. Fernando AGUILAR
Université de Neuchâtel
Laboratoire de Physiologie Végétale et Biochimie
Case postale 2
CH - 2007 NEUCHATEL
Switzerland

Dr. ARGYROUDI-AKOYOUNOGLOU
Demokritos Research Center
Aghia Paraskevi
ATTIKA
Greece

Dr. Juan P. G. BALLESTA
CSIC and UAM
Universidad Autonoma de Madrid
Cantoblanco
28049 MADRID
Spain

Mr. Christophe BONNY
Université de Neuchâtel
Laboratoire de Biochimie Végétale
Case postale 2
CH - 2007 NEUCHATEL
Switzerland

Dr. A. BOSCHETTI
Institut für Biochemie
Universität Bern
Freiestrasse 3
CH-3012 BERN
Switzerland

Dr. Don BOURQUE
Dept of Biochemistry
University of Arizona
TUCSON, Arizona 85721
USA

Dr. Catherine M. BOWMAN
IPSR Cambridge laboratory
Centre for Plaht Science Research
Colney Lane
Norwich NR4 7UJ
G-B

Dr. R. BRIMACOMBE
Max-Planck Institut
für Molekulare Genetik
Ihnestrasse 73
D-1000 BERLIN 33 (DAHLEM)
FRG

Dr. ÇIRAKOGLU
Research Institute for Basic Sciences
Tubitak
GEBZE / KOCAELI
B. O. Box 74
Turkey

Mr. Pierre CAROL
Laboratoire de Biologie Moléculaire Végétale
Université Joseph Fourier
B.P. 53X
38041 GRENOBLE CEDEX
France

Dr. B. EHRESMANN
Institut de Biologie Moléculaire et Cellulaire
15 rue Descartes
67084 STRASBOURG CEDEX
France

Dr. J. FEIERABEND
Botanisches Institut
J.W. Goethe-Universität
Postf. 111932
D-6000 FRANKFURT A. MAIN
FRG

Dr. S. GANTT
Dept of Plant Biology
University of Minnesota
1445 Gortner Ave
SAINT PAUL, Minnesota 55108
USA

Dr. K. GAUSING
Department of Molecular Biology
and Plant Physiology
University of Aarhus
DK-8000 ARHUS C
Denmark

Dr.Giselle GEDDES
University of Aberdeen
Department of Molecular and cell Biology
Marischal College
ABERDEEN AB9 1AS
Great-Britain

Dr. K. GIESE
Max-Planck Institut
für Molekulare Genetik
Ihnestrasse 63-73
D-1000 BERLIN 33 - FRG

Dr. N. W. GILLHAM
Dept. of Botany & Zoology
Duke University
DURHAM, NC 27706
USA

Dr. R.B. HALLICK
Dept of Biochemistry
University of Arizona
TUCSON, Arizona 85721
USA

Dr. HARRIS
Dept. of Botany & Zoology
Duke University
DURHAM, NC 27706
USA

Dr. C. JOHNSON
Max-Planck Institut für Molekulare Genetik
Ihnestrasse 63-73
D-1000 BERLIN 33
FRG

Dr. P.H. van KNIPPENBERG
Department of Biochemistry
State University of Leiden
P.O. Box 9502
2300 RA LEIDEN
Netherlands

Dr. H. KÖSSEL
Institut für Biologie III
Schänzle strasse 1
D-7800 FREIBURG
FRG

Dr. Marcel KUNTZ
IBMP
12 rue du Général Zimmer
67084 STRASBOURG
France

Mr. Thierry LAGRANGE
Laboratoire de Biologie Moléculaire Végétale
Université Joseph Fourier
B.P. 53X
38041 GRENOBLE CEDEX
France

Dr. Ester LAZARO
Centro de Biologia Molecular
Universidad Autonoma de Madrid
Campus de Cantoblanco
28049 MADRID - Spain

Dr. W. LÖFFELHARDT
Institut für Allgemeine Biochemie
und Ludwig-Boltzmann-Forschungstelle für
Biochemie
Universität Wien
Wäringerstrasse 38
A-1090 WIEN
Austria

Dr. S. LOISEAUX-de GOER
Station Biologique de Roscoff
B.P. 74
29682 ROSCOFF CEDEX
France

Dr. R. MACHE
Laboratoire de Biologie Moléculaire Végétale
Université Joseph Fourier
B.P. 53X
38041 GRENOBLE CEDEX
France

Dr. Roland MAY
Institut Laue-Langevin
B.P. 156X
38042 GRENOBLE CEDEX
France

Dr R. MATAGNE
Genetics of Microorganisms
Department of Botany
University of Liège
Sart Tilman
B-4000 LIEGE
Belgium

Dr. B. RANDOLPH-ANDERSON
Dept. of Botany & Zoology
Duke University
DURHAM, NC 27706
USA

Dr. J.D. ROCHAIX
Département de Biologie Moléculaire
Université de Genève
30 quai E. Ansermet
21211 GENEVE 4,
Switzerland

Dr. Pascale ROMBY
Institut de Biologie
Moléculaire et Cellulaire
15 rue Descartes
67084 STRASBOURG
France

Dr. SPREMULLI
Dept of Chemistry
University of North Carolina
CHAPEL HILL, NC 27599-3290
USA

Dr. John STEPHEN
University of Aberdeen
Department of Molecular and cell Biology
Marischal College
ABERDEEN AB9 1AS
Great-Britain

Dr. E. STUTZ
Université de Neuchâtel
Laboratoire de Biochimie Végétale
Case postale 2
CH - 2007 NEUCHATEL
Switzerland

Dr. A. R. SUBRAMANIAN
Max-Planck Institut für Molekulare Genetik
Abteilung Wittmann
D-1000 BERLIN DAHLEM
RFA

Dr. M. SUGIURA
Center for Gene Research
Nagoya University
NAGOYA 464-01
Japan

Dr. M. THOMPSON
Dept of Plant Biology
University of Minnesota
1445 Gortner Ave
SAINT PAUL, Minnesota 55108
USA

Dr. J. H. WEIL
Institut de Biologie Moléculaire des Plantes du CNRS
Université Louis Pasteur
12 rue du Général Zimmer
67084 Strasbourg cedex
France

SUBJECT INDEX

NATO ASI Series H

Vol. 1: **Biology and Molecular Biology of Plant-Pathogen Interactions.** Edited by J. A. Bailey. 415 pages. 1986.

Vol. 2: **Glial-Neuronal Communication in Development and Regeneration.** Edited by H. H. Althaus and W. Seifert. 865 pages. 1987.

Vol. 3: **Nicotinic Acetylcholine Receptor: Structure and Function.** Edited by A. Maelicke. 489 pages. 1986.

Vol. 4: **Recognition in Microbe-Plant Symbiotic and Pathogenic Interactions.** Edited by B. Lugtenberg. 449 pages. 1986.

Vol. 5: **Mesenchymal-Epithelial Interactions in Neural Development.** Edited by J. R. Wolff, J. Sievers, and M. Berry. 428 pages. 1987.

Vol. 6: **Molecular Mechanisms of Desensitization to Signal Molecules.** Edited by T. M. Konijn, P. J. M. Van Haastert, H. Van der Starre, H. Van der Wel, and M. D. Houslay. 336 pages. 1987.

Vol. 7: **Gangliosides and Modulation of Neuronal Functions.** Edited by H. Rahmann. 647 pages. 1987.

Vol. 8: **Molecular and Cellular Aspects of Erythropoietin and Erythropoiesis.** Edited by I. N. Rich. 460 pages. 1987.

Vol. 9: **Modification of Cell to Cell Signals During Normal and Pathological Aging.** Edited by S. Govoni and F. Battaini. 297 pages. 1987.

Vol. 10: **Plant Hormone Receptors.** Edited by D. Klämbt. 319 pages. 1987.

Vol. 11: **Host-Parasite Cellular and Molecular Interactions in Protozoal Infections.** Edited by K.-P. Chang and D. Snary. 425 pages. 1987.

Vol. 12: **The Cell Surface in Signal Transduction.** Edited by E. Wagner, H. Greppin, and B. Millet. 243 pages. 1987.

Vol. 13: **Toxicology of Pesticides: Experimental, Clinical and Regulatory Perspectives.** Edited by L. G. Costa, C. L. Galli, and S. D. Murphy. 320 pages. 1987.

Vol. 14: **Genetics of Translation. New Approaches.** Edited by M. F. Tuite, M. Picard, and M. Bolotin-Fukuhara. 524 pages. 1988.

Vol. 15: **Photosensitisation. Molecular, Cellular and Medical Aspects.** Edited by G. Moreno, R. H. Pottier, and T. G. Truscott. 521 pages. 1988.

Vol. 16: **Membrane Biogenesis.** Edited by J. A. F. Op den Kamp. 477 pages. 1988.

Vol. 17: **Cell to Cell Signals in Plant, Animal and Microbial Symbiosis.** Edited by S. Scannerini, D. Smith, P. Bonfante-Fasolo, and V. Gianinazzi-Pearson. 414 pages. 1988.

Vol. 18: **Plant Cell Biotechnology.** Edited by M. S. S. Pais, F. Mavituna, and J. M. Novais. 500 pages. 1988.

Vol. 19: **Modulation of Synaptic Transmission and Plasticity in Nervous Systems.** Edited by G. Hertting and H.-C. Spatz. 457 pages. 1988.

Vol. 20: **Amino Acid Availability and Brain Function in Health and Disease.** Edited by G. Huether. 487 pages. 1988.

NATO ASI Series H

Vol. 21: **Cellular and Molecular Basis of Synaptic Transmission.**
Edited by H. Zimmermann. 547 pages. 1988.

Vol. 22: **Neural Development and Regeneration. Cellular and Molecular Aspects.**
Edited by A. Gorio, J.R. Perez-Polo, J. de Vellis, and B. Haber. 711 pages. 1988.

Vol. 23: **The Semiotics of Cellular Communication in the Immune System.**
Edited by E.E. Sercarz, F. Celada, N.A. Mitchison, and T. Tada. 326 pages. 1988.

Vol. 24: **Bacteria, Complement and the Phagocytic Cell.**
Edited by F.C. Cabello und C. Pruzzo. 372 pages. 1988.

Vol. 25: **Nicotinic Acetylcholine Receptors in the Nervous System.**
Edited by F. Clementi, C. Gotti, and E. Sher. 424 pages. 1988.

Vol. 26: **Cell to Cell Signals in Mammalian Development.**
Edited by S.W. de Laat, J.G. Bluemink, and C.L. Mummery. 322 pages. 1989.

Vol. 27: **Phytotoxins and Plant Pathogenesis.**
Edited by A. Graniti, R.D. Durbin, and A. Ballio. 508 pages. 1989.

Vol. 28: **Vascular Wilt Diseases of Plants. Basic Studies and Control.**
Edited by E.C. Tjamos and C.H. Beckman. 590 pages. 1989.

Vol. 29: **Receptors, Membrane Transport and Signal Transduction.**
Edited by A.E. Evangelopoulos, J.P. Changeux, L. Packer, T.G. Sotiroudis, and K.W.A. Wirtz. 387 pages. 1989.

Vol. 30: **Effects of Mineral Dusts on Cells.**
Edited by B.T. Mossman and R.O. Bégin. 470 pages. 1989.

Vol. 31: **Neurobiology of the Inner Retina.**
Edited by R. Weiler and N.N. Osborne. 529 pages. 1989.

Vol. 32: **Molecular Biology of Neuroreceptors and Ion Channels.**
Edited by A. Maelicke. 675 pages. 1989.

Vol. 33: **Regulatory Mechanisms of Neuron to Vessel Communication in Brain.**
Edited by F. Battaini, S. Govoni, M.S. Magnoni, and M. Trabucchi. 416 pages. 1989.

Vol. 34: **Vectors as Tools for the Study of Normal and Abnormal Growth and Differentiation.**
Edited by H. Lother, R. Dernick, and W. Ostertag. 477 pages. 1989.

Vol. 35: **Cell Separation in Plants: Physiology, Biochemistry and Molecular Biology.**
Edited by D.J. Osborne and M.B. Jackson. 449 pages. 1989.

Vol. 36: **Signal Molecules in Plants and Plant-Microbe Interactions.**
Edited by B.J.J. Lugtenberg. 425 pages. 1989.

Vol. 37: **Tin-Based Antitumour Drugs.** Edited by M. Gielen. 226 pages. 1990.

Vol. 38: **The Molecular Biology of Autoimmune Disease.**
Edited by A.G. Demaine, J-P. Banga, and A.M. McGregor. 404 pages. 1990.

Vol. 39: **Chemosensory Information Processing.** Edited by D. Schild. 403 pages. 1990.

Vol. 40: **Dynamics and Biogenesis of Membranes.**
Edited by J.A.F. Op den Kamp. 367 pages. 1990.

Vol. 41: **Recognition and Response in Plant-Virus Interactions.**
Edited by R.S.S. Fraser. 467 pages. 1990.

NATO ASI Series H

Vol. 42: **Biomechanics of Active Movement and Deformation of Cells.**
Edited by N. Akkaş. 524 pages. 1990.

Vol. 43: **Cellular and Molecular Biology of Myelination.**
Edited by G. Jeserich, H. H. Althaus, and T. V. Waehneldt. 565 pages. 1990.

Vol. 44: **Activation and Desensitization of Transducing Pathways.**
Edited by T. M. Konijn, M. D. Houslay, and P. J. M. Van Haastert. 336 pages. 1990.

Vol. 45: **Mechanism of Fertilization: Plants to Humans.**
Edited by B. Dale. 710 pages. 1990.

Vol. 46: **Parallels in Cell to Cell Junctions in Plants and Animals.**
Edited by A. W. Robards, W. J. Lucas, J. D. Pitts, H. J. Jongsma, and D. C. Spray. 296 pages. 1990.

Vol. 47: **Signal Perception and Transduction in Higher Plants.**
Edited by R. Ranjeva and A. M. Boudet. 357 pages. 1990.

Vol. 48: **Calcium Transport and Intracellular Calcium Homeostasis.**
Edited by D. Pansu and F. Bronner. 456 pages. 1990.

Vol. 49: **Post-Transcriptional Control of Gene Expression.**
Edited by J. E. G. McCarthy and M. F. Tuite. 671 pages. 1990.

Vol. 50: **Phytochrome Properties and Biological Action.**
Edited by B. Thomas and C. B. Johnson. 337 pages. 1991.

Vol. 51: **Cell to Cell Signals in Plants and Animals.**
Edited by V. Neuhoff and J. Friend. 404 pages. 1991.

Vol. 52: **Biological Signal Transduction.**
Edited by E. M. Ross and K. W. A. Wirtz. 560 pages. 1991.

Vol. 53: **Fungal Cell Wall and Immune Response.**
Edited by J. P. Latgé and D. Boucias. 472 pages. 1991.

Vol. 54: **The Early Effects of Radiation on DNA.**
Edited by E. M. Fielden and P. O'Neill. 448 pages. 1991.

Vol. 55: **The Translational Apparatus of Photosynthetic Organelles.**
Edited by R. Mache, E. Stutz, and A. R. Subramanian. 260 pages. 1991.

Vol. 56: **Cellular Regulation by Protein Phosphorylation.**
Edited by L. M. G. Heilmeyer, Jr. 520 pages. 1991.

Vol. 57: **Molecular Techniques in Taxonomy.**
Edited by G. M. Hewitt, A. Johnston, and J. P. W. Young. 420 pages. 1991.

MIX
Papier aus verantwortungsvollen Quellen
Paper from responsible sources
FSC® C105338

If you have any concerns about our products,
you can contact us on
ProductSafety@springernature.com

In case Publisher is established outside the EU,
the EU authorized representative is:
Springer Nature Customer Service Center GmbH
Europaplatz 3, 69115 Heidelberg, Germany

Printed by Libri Plureos GmbH
in Hamburg, Germany